Dis ce que tu sais, fais ce que dois, advienne que pourra.

Sofia Kovalevskaya (1888)

MATHEMATICIAN WITH THE SOUL OF A POET

POEMS AND PLAYS
OF
SOFIA KOVALEVSKAYA

TRANSLATED, EDITED AND INTRODUCED BY
SANDRA DELOZIER COLEMAN

BOHANNON HALL PRESS

Cover art designed by Sandra D. Coleman from an Isaak Brodsky
painting[1] and a photo illustrating the history of hydropower.

Library of Congress Control Number: 2021919995

Publisher's Cataloging-in-Publication Data

Coleman, Sandra DeLozier, 1949-
    Mathematician with the soul of a poet: Poems and plays of Sofia Kovalevskaya /
    translated by Sandra DeLozier Coleman.

    Translation of The Struggle for Happiness, by S. V, Kovalevskaya and A. C.
    Leffler. Beauty is Earth's Sacred Secret, by Karl Weierstrass.
    p. cm. Includes bibliographical references.

1. Women mathematicians—19th century—Literary collections. 2. Swedish literature-
-Translations into English—History and criticism. I. Title. II. Kovalevskaya, S.V.,
1850-1891—Translations into English. III. Weierstrass, Karl, 1815-1897—Translations
into English. IV. Leffler, Anne Charlotte, 1849-1892—Translations into English.
PT9567.C65 M38 2021   2021919995

ISBN 979-8-9850298-0-2 (softcover)
ISBN 979-8-9850298-1-9 (hardcover)

Published by Bohannon Hall Press

With love to Stephanie and Sydney
who are enjoying
the freedom
to study science and mathematics
and whatsoever else they may choose

TABLE OF CONTENTS

PREFACE

During the twelve years I served as Book Review Editor for *The AMATYC Review*,[2] I read and reviewed many interesting books, but only one group inspired me in a way that had a strong and lasting effect on my life. After reviewing three books on Sofia Kovalevskaya and learning that, in addition to being celebrated for her significant accomplishments in mathematics, she had been a prolific and successful writer, I found myself wanting to learn more about her writings. I became particularly interested in the content of her nine poems and the two plays I had read about in nearly every account of Sofia Kovalevskaya's life.

This book is not intended to be another biography of Sofia Kovalevskaya. Several excellent biographies have already been written. References to Kovalevskaya's life events and personality traits are scattered throughout the book in commentary, brief discussions and even within the endnotes, but readers should understand that these references provide only glimpses into a much larger story. Those discovering Sofia Kovalevskaya for the first time through the poems and plays will want to learn more about her remarkable life by reading at least one of the several existing biographies or by reading Joan Spicci's entertaining and informative historical novel, *Beyond the Limit*.[3]

Since a brief introduction to her life can be found in the set of three reviews I wrote for *The AMATYC Review* in the fall of 1996, I decided it might be helpful to begin by sharing that review in full as it appeared in the two-year college math journal.[4] These were the first of many reviews of the kinds of math-related books that could appeal to a broad audience, not just to people who have studied mathematics. My review articles were published twice a year and were generally several pages long. Most tied two or three related books together and sometimes linked the books to personal life experiences. A number of editions of the journal included poems I had written related to the theme of the group of books in the review or to other math topics.

The Kovalevskaya books and the reviews of them were just the beginning of an adventure that has lasted more than twenty years. They tell of my personal introduction to her story and mark the beginning of my long search for the image of Sofia Kovalevskaya as a mathematician with the soul of a poet.

Reading the three Kovalevskaya books opened to me a part of history I had not known, but should have. My library now includes these and other Kovalevskaya biographies, many other books important to her story, a huge collection of SK articles and a group of posters I once made of images of key people in her life. I have acquired copies of significant papers, including her doctoral dissertation and her Prix Bordin submission, and have amassed a substantial collection of photographs of places where she lived or visited: St. Petersburg, Heidelberg, Berlin, Beaulieu-sur-Mer, Nice, London, and Paris. Tracing her steps as I have traveled has been a great joy to me and a source of many treasured memories! Next stop: Stockholm and the Institute Mittag-Leffler! Who knows? Someday I may satisfy the long-time dream of visiting her Palibino childhood home.

Book Reviews
Edited by Sandra DeLozier Coleman

A CONVERGENCE OF LIVES, Ann Hibner Koblitz, Rutgers University Press, 1993, xxxviii + 305 pages, ISBN 0-8135-1962-4 / 0-8135-1963-2 pbk.

A RUSSIAN CHILDHOOD, Sofya Kovalevskaya, translated by Beatrice Stillman, Springer-Verlag, 1978, xiii + 250 pages, ISBN 0-387-90348-8.

LITTLE SPARROW, Don H. Kennedy, Ohio University Press, 1983, ix + 341 pages, ISBN 0-8214-0692-2 / 0-8214-0703-1 pbk.

One afternoon last year, I received a postcard from Rutgers University Press advertising a book by Ann Hibner Koblitz about the life of Sofia Kovalevskaya.[5] The book is called *A Convergence of Lives,* because it brings to light many different facets of the life of this remarkable woman. An entire book devoted to the life of one woman mathematician seemed to me to be rare, even in today's world, as Kovalevskaya's presence was in the university environment of the late 19th century. I was intrigued and ordered the book immediately. It arrived shortly, and, as I began to read, I found the account of Kovalevskaya's success in establishing herself as a respected colleague among the many famous European mathematicians of her day to be both fascinating and inspiring.

She was born in Russia in 1850, at a time when the world of mathematics in her homeland was closed to all women regardless of interests or talents. At the age of eighteen, Sofia Korvin-Krukovskaia's desire to find a place where she could further her education was so great that she entered into a fictitious marriage with a young nihilist, Vladimir Kovalevsky, who agreed to help her travel to Germany and to leave her there to study mathematics. Though frowned upon by parents, such marriages were not unusual among the young revolutionaries of Russia who were convinced that furthering education, especially in the sciences, was the key to conquering all of the ills of society.

In *A Convergence of Lives,* Koblitz brings together images of Kovalevskaya as an idealist, a student, a teacher, a writer, a scientist, a poet, a mathematician and an editor of one of the world's most respected mathematics research journals. Koblitz also presents a very personal image of a child, a sister, a friend, a wife and a mother. We follow Kovalevskaya's footsteps as she boldly introduces herself to the famous German mathematician, Karl Weierstrass, and convinces him that she is capable of learning all the mathematics he has time to teach her.

We share her excitement as she earns the distinction of being the first woman to be awarded a PhD in Mathematics; we feel her frustration as she strives to obtain a satisfying teaching position in higher mathematics; and we enjoy her triumph as she goes on to win the Prix Bordin, an honor as great as winning the Nobel Prize.[6]

Koblitz's account makes us so keenly aware of the number and intensity of the obstacles Sofia Kovalevskaya faced that men and women in the field of mathematics who read her story are sure to be inspired by the fact that she prevailed against such odds.

Although the Koblitz biography contains more information about Sofia Kovalevskaya than is generally available about most famous mathematicians, I soon discovered, by exploring the suggested options for further reading, that it is possible to become so well acquainted with Kovalevskaya's life as to feel one has known her personally.

Her autobiographical book, *A Russian Childhood,* translated by Beatrice Stillman and published by Springer-Verlag in 1978, gives a moving account of her early childhood and adolescence and provides much insight into the psychological and personal aspects of the life of this developing mathematician. I know of no other mathematician who has given the world such a detailed account of childhood experiences. The book is enjoyable both for the stories of significant events in Kovalevskaya's life and for the enlightening portrayal of aspects of 19th century Russian society.

During her lifetime, Kovalevskaya received as much acclaim for her literary accomplishments as for her mathematics. She began composing poetry at the age of five and wrote poetry all her life. This may have been one of the things Weierstrass admired about her, since he once wrote to her saying, "It is completely impossible to be a great mathematician without having the soul of a poet." She also wrote literary, science and theatrical reviews, as well as, novels and plays – some of which have been published in many languages – most notably the memoir, *A Russian Childhood.*

In *A Russian Childhood,* beginning with the day she first learned to say her name, Kovalevskaya unfolds the story of her life with such a passionate desire to have the world know who she was and what she was like that the reader soon begins to see her, even in the glory of her mathematical success, as just little "Sonya" all grown up! She holds nothing back – freely sharing her joys, her sorrows, her fears, and even her adolescent love for the great writer and family friend, Fyodor Dostoevsky. She ends her story with the end of that relationship and leaves the reader longing to know what happens next.

Fortunately, Don H. Kennedy in his book, *Little Sparrow,* fills in more of the details and leaves the reader with a more complete understanding of Kovalevskaya's experiences and personality. Dedicated to the author's wife because of her untiring efforts to locate and translate Russian books, periodicals, letters, and other original documents related to Kovalevskaya's life, his book contains recollections of stories about "Sonya" handed down to family members of those who knew her personally.

4

Here, the reader comes to understand her strengths as well as her weaknesses. We are given a clearer picture of what her life was like during the many years during which she and her husband maintained a fictitious marriage. We learn why they decided to change that arrangement. The five years during which Kovalevskaya gave up mathematics almost entirely are explained in some detail. Her long-standing relationship with the famous German mathematician Karl Weierstrass is presented in such a way as to be an inspiration to both aspiring mathematicians and their mentors.

As one of the editors of the highly respected mathematics journal, *Acta Mathematica*, Kovalevskaya was familiar with all of the current developments in mathematics and corresponded with mathematicians throughout Europe. Her life is so intimately intertwined with the lives of such well-known scientists and mathematicians as Darwin, Bunsen, Mittag-Leffler, Kronecker, Cantor, Hermite, Poincaré, Picard, Chebyshev, and Sylvester that reading her story sheds light on the whole picture of the interrelationships that played a role in the development of modern mathematics in the 19th century.

As in the other two books, photographs of Kovalevskaya and of those who played prominent roles in her life give the story a completeness I have not found in the accounts of the lives of other mathematicians. Details concerning her literary and mathematical contributions are included in an appendix.

Among other accomplishments, she developed the final form of the Cauchy-Kovalevskaya Theorem for partial differentiation, solved a problem related to the revolution of a solid body about a fixed point that had baffled mathematicians for over one hundred years and wrote a paper related to refraction of light in a crystalline medium.

Through these biographical accounts, Sofia Kovalevskaya comes to life for the reader; but what is probably more significant she is revealed as a person with whom the reader may identify. As this happens, her successes become inspiration. We see her face her weaknesses, some of which she overcomes, but some of which she matter-of-factly accepts, as she continues to move forward to achieve her ultimate goal – to make a difference in the world – to make a lasting and significant contribution.

Reviewed by the Editor, Sandra DeLozier Coleman

So, a story that began with a book review has led to a book! Many people helped to make this book possible, whether through promoting my plans, obtaining materials, refining the translations, careful proofing, or publication. Every contribution is appreciated. Those who read and made suggestions on the manuscript or provided attentive audience as I read passage after passage from the plays, or who attended readings of the poems in the US or Russia – all gave me encouragement as I pursued my dream.

I would like to express special appreciation to: Michael Dutko, John and Diane Konvalina, Charles Seife, Anna Klein, Larisa Galperina, Frank Richards, Dmitriy Leykekhman, Joan Spicci, JoAnne Growney, Elena Novikova, Kjell-Owe Widman, Natalia Krasilnikova, Claire and Helaman Ferguson, Janice Henderson, Alysa Salzberg, Tom Blair, Sheila Patrick, Sharon Woodman, Sarah Coleman and Jennifer Caddell.

I would also like to add a note of gratitude to all of the researchers, biographers and translators whose efforts have served to acquaint readers with the life and works of Sofia Kovalevskaya and to help keep her story alive.

There are not enough words to express my deep appreciation to my husband, Rip Coleman, for the encouragement, love, help, and support he has given me. From listening to me read the first review to the final editing of this book, he has been devotedly sharing this wonderful adventure every step of the way.

SEARCHING FOR THE SOUL OF A POET

I understand that you are surprised that I can work at the same time in both literature and mathematics. Many who have not had the chance to learn more about mathematics confuse it with arithmetic and consider it to be a dry and arid science. In truth, however, this science requires the greatest imagination, and one of the most respected mathematicians of our century has very rightly said that it is not possible to be a great mathematician without having the soul of a poet.

S. V. Kovalevskaya

Sofia Vasilievna Kovalevskaya expressed these thoughts in a letter to Russian writer, Aleksandra Stanislavovna Shabelskaya,[7] in the autumn of 1890. The publication of Kovalevskaya's autobiographical book, *A Russian Childhood*,[8] had caused quite a stir in literary circles earlier that year, being praised by some as comparable in quality to similar works by Tolstoy and Turgenev. Already celebrated and respected as the first woman since Renaissance Italy to earn a doctoral degree in mathematics, Kovalevskaya had recently been proclaimed winner of the prestigious Prix Bordin award for her elegant presentation of a long-sought solution to the famous *mermaid* problem related to the motion of a solid body about a fixed point. She was in correspondence, as an equal, with all of the most productive mathematicians in Europe. She was well-known as an editor of the prominent journal, *Acta Mathematica*, and is remembered today for a number of significant mathematical results. Yet, in spite of her clear success in the field of mathematics, it seems that, because of the praise she was receiving for her recent literary publication, some were speculating that Kovalevskaya would abandon mathematics altogether for what they considered to be the more feminine occupation of writing.

Kovalevskaya seemed to be speaking in her own defense when she declared the fields of writing and mathematics to be more complementary than contradictory in nature, quoting her dear friend and mentor, Karl Weierstrass, in asserting that a great mathematician must have something of the soul of a poet. It is apparent from the comments in her letter to Shabelskaya that she would have gone on working in both fields indefinitely had a sudden illness during the following winter not ended her life prematurely. She had written in her letter that "the poet has only to perceive what others do not perceive, to look deeper than others look. And the mathematician must do the same thing." [9] It is also clear from her reference to his words that Kovalevskaya understood that Weierstrass thought of her as a mathematician with the soul of a poet.

The acclaimed autobiographical story of her childhood has earned Kovalevskaya much praise for her writing. It provides a clear and colorful account of her emotional and intellectual development as well as an accurate picture of Russian family life as she experienced it. Thanks to an excellent translation and introduction by Beatrice Stillman, published by Springer-Verlag in 1978, English-speaking readers are able to come to know Sofia Kovalevskaya through her own words. *A Russian Childhood* is an engaging story. People interested in her life will read it for many years to come and will be moved by how effectively she was able to convey images of her innermost self, as she confessed feelings as intimate as her intense adolescent love for the writer and family friend, Fyodor Dostoevsky, and even her sincere prayer to be pretty, even if only in his eyes.

Sofia Kovalevskaya's chief aim in all her writing seems to have been to have the world know her well and to remember her, not just as a mathematician, but as a person of deep feeling and broad interests. The novels, essays, letters, diary entries, scientific articles, theatrical reviews, plays, and poems, through which she shared her thoughts on everything from politics to the existence of soulmates, all serve to reveal aspects of her personality that are of value, regardless of their literary merit. Collectively, they reveal an individual whose life included wide-ranging interests and multiple aspirations – a person who wanted very much to be seen as such and who resisted being confined in any way to any developing narrow views of a woman mathematician.

The more we learn about Sofia Kovalevskaya through her writings, the better we see that she paved the way for women to pursue the study of mathematics without impediment in more ways than one. By holding fast to her right to pursue other interests, to develop other talents and to reveal her thoughts and feelings in many different forms of writing, she made apparent the important truth that there is no confining box into which a woman mathematician must fit.

This could be considered almost as significant an offering to the world of mathematics as was her mathematical research. Too many mathematicians, especially women, know the feeling of being summed up and dismissed by people who expect to have nothing in common with any person introduced as a mathematician. Sharing knowledge about the very human side of a celebrated, successful mathematician helps dispel such limiting ideas and opens pathways for making connections on many levels – a not insignificant factor in encouraging mathematically talented young women faced with making important life choices to make the choice to pursue mathematics.

There has been much difference of opinion among biographers related to what aspects of Sofia Kovalevskaya's personality should be presented or emphasized when writing or speaking about her. One biographer in discussing Kovalevskaya's efforts at poetry has dismissed her poems as of "marginal literary interest,"[10] saying that her "point of view [is] too naïve and sentimental to appeal to modern tastes." Another has said that her poems were "only verses written by one without a true gift for poetic expression" adding that "it is no loss if all are forgotten."[11]

While it is doubtful that Sofia Kovalevskaya's small collection of poems formed the basis for Karl Weierstrass' view of his friend and protégé as a mathematician with the soul of a poet, it is probable that her writing contributed much to that image. During their more than twenty years of correspondence, he would have become aware of her perspective on many aspects of life – not only through her letters to him, but also through her other writings. It is certainly not because she wrote poems having universal appeal that Weierstrass saw Sofia Kovalevskaya as a mathematician with the soul of a poet. He had an appreciation for the imagination, creativity and passion she brought to her mathematical pursuits, to her writings, and to her personal life.

It is, of course, essential in writing about Kovalevskaya to tell of her well-deserved standing as a prominent mathematician. She was a highly respected member of an elite group of important mathematicians in 19th century Europe. It is also important to note that her work on differential equations continues to influence the research of mathematicians around the world.[12] It is understandable that biographers would want to correct the record where writers or speakers, past or present, may have diminished the importance of her work or spread false information about her. Where writers have included in biographies stories related to highly-opinionated early criticisms of Kovalevskaya, we could make a case for omitting such unfounded criticisms or disregarding them as insignificant without losing much of her story.

We do lose much, however, if we omit as insignificant what Kovalevskaya revealed about herself through her poetry and other writings. To totally discount these sources of biographical information implies that to take her mathematics seriously, we must ignore, or even hide, many of the thoughts she sought to share and some of her personality traits. This does her a disservice and could also actually discourage young women of today who might be considering becoming mathematicians.

There are still many talented young women who would pursue the study of mathematics were it not for the fear, especially in adolescent years, of being put into an undesirable box built of inappropriate labels and false assumptions regarding the personalities of women mathematicians. Sofia Kovalevskaya's story is unique in its details, but it need not be unique as an example of a woman mathematician whose life had many facets. A young mathematician does not need to have a personality like Sofia Kovalevskaya's or a dramatic life story like hers in order to feel free to explore multiple interests, to live a full and interesting life or to share personal accounts of varied life experiences. Sofia Kovalevskaya's successes in the field of mathematics are evidence that pursuing thoughts or experiences far removed from science does not erase the ability to learn mathematics or the capability of offering new ideas to the field.

A strong belief that Kovalevskaya's story encourages personal freedom for budding mathematicians was behind my efforts to translate the poems and plays included in this book. These writings reveal much about one woman mathematician's spirit, hopes, disappointments, and ideals. We can try to assess how certain events and associated feelings may have helped or hindered her studies. But, in any case, they are part of her story. I do not believe they should be forgotten and I am pleased to be able to offer English translations so that more readers who, like myself, have wanted very much to know the content of the poems and plays can now have easy access.

Because of the intensity of my own interest, I can certainly imagine an interested audience for the Kovalevskaya poems and plays. When I learned about Sofia Kovalevskaya through the books I read and reviewed in 1996 as Book Review Editor[13] for *The AMATYC Review* and discovered that she had written poems, I found myself obsessed with wanting to know more about them.

From childhood, I had also pursued interests in both writing and mathematics equally, so I may have had a special curiosity about her dual zeal for both.[14] I, too, had written poems as early as age five, had won a small prize for poetry at age ten and had been inducted into the literary honorary, *Quill and Scroll*,[15] in high school at about the same time I became active in math competitions and the high school math club. I went on to earn degrees in mathematics and continued with various forms of writing. I wrote over one thousand poems, engaged in long and significant correspondences and regularly published book reviews.

Hoping to learn more about Sofia Kovalevskaya from her poems and plays, I was very much disappointed to find that there were no English translations. A friend tried to help by sending a box filled with the 560⁺ pages of *Vospominaniya Povesti (VP)*, the complete collection of all of Sofia Kovalevskaya's writings. He had hinted that a special Christmas gift was coming in an e-mail headed "How's Your Russian?"

I knew no Russian and had no knowledge of the existence of *VP*, so I was ecstatically surprised by my gift when it arrived. The friend, who had met me when I was a student in his classes, had been corresponding with me since I had completed my master's degree and knew through e-mail of my interest in finding the poems. He had enlisted the help of his wife, who had worked at the university library during her student years, in locating the book of writings and arranging to have it copied. The *VP* gift box they sent me was filled with legal-sized copies of pages and pages of Russian. All of Sofia Kovalevskaya's writings were right in front of me. If only I could speak or read Russian!

Locating the poems among the loose pages was not a problem, because of their structure. Once located, I immediately rushed to Books-a-Million to purchase a Russian/English dictionary. At home, I began trying to translate the first poem one word at a time. This proved to be a particularly slow and difficult process for someone who was not yet even familiar with the order or sounds of the letters of the Russian alphabet. Nevertheless, because I wanted so much to know the content of the poems, it was worth the effort to learn.

At the bookstore, I had also purchased the first of many sets of tapes to help me begin to teach myself Russian. A set produced by Passport Books titled *Just listen 'n' Learn Russian: The Basic Course for Succeeding in Russian and Communicating with Confidence* suggested that I should be able to reach those lofty goals with just the very small book and three cassette tapes included! I was not so naïve as to believe it would be that easy, but it was a start! Of course, travel dialog does little to develop poetic vocabulary, so, for the translations, I was prepared to make extensive use of the excellent 1000-page dictionary I had found. Kenneth Katzner's *English-Russian/Russian-English Dictionary*, published by John Wiley & Sons, is an amazingly complete resource with all the various usages of nearly every Russian word presented in the context of complete phrases. It is a very useful tool for translating and learning Russian.

By the time of the MAA/AMS Joint Mathematics Meetings,[16] in January of 1997, I was confident enough in my translation of *Chameleon* to present the poem during the annual *Evening of Poetry*. The event is an informal gathering of mathematicians who write poetry or share poems written by mathematicians. There, I explained how much I was hoping to translate the Kovalevskaya poems but how difficult it was going to be for me, since I was only beginning to learn Russian. At the time, there was no available translation software and I had no Russian-speaking companions.

A few days after the conference, I received an e-mail from Michael Dutko, a mathematician who had been in the audience. He told me he was a math professor at Scranton University and that his native language was Russian. He wrote that he would be pleased to provide the initial word for word meaning of the poems. He had heard my presentations of several of my own poems at the poetry gathering and he trusted that I would be able to use what he could provide to develop effective translations in poetic form.

I eagerly accepted his offer and we began our collaboration. I was teaching mathematics at the Avery Point campus of the University of Connecticut. He was teaching at Scranton University in Pennsylvania. So, we had to do most of our work through e-mail. We met at my home in Groton Long Point a couple of times, in New York once and once somewhere in between our universities. We worked on nine poems for nine months, completing all but one during that time.[17]

During those nine months of working on the poems, I was also making great efforts to learn to speak Russian. In addition to listening to Russian language tapes, I began to buy every book I could find to help me learn – new and used. I approached and tried to make friends with anyone I overheard speaking with a Russian accent, whether out shopping, dining, or at the university. I began asking Russian-speaking business associates of my husband to let me record them reading some of the Kovalevskaya poems.

Before long, everyone I knew had heard that I was trying to teach myself Russian, including my pastor's wife. She decided to give me an assignment. She envisioned a Pentecost Sunday event in which as many people as she could enlist would begin to recite John 3:16 together in as many languages as members of our congregation could provide. I was asked to learn to recite the passage in Russian.

Michael Dutko was visiting my home in Groton Long Point when I asked him to help me say the words correctly. These were the first words in Russian that I learned to say by heart. I attached the words to a moving image of Jesus that my sister had drawn during a troubled period of her life. I printed the image along with the Russian words and posted the page on my refrigerator door where it remained for many months.

Groton Long Point was a very unusual place to live. Most of the buildings, including our home, were vacation homes occupied primarily in the summer either by owners or by beach visitors willing to pay high prices to be near the beach for a week or two.

We had made a special arrangement with the owners of the six-bedroom home where we lived. We rented the huge furnished house for a low price during the off season while I taught at the Avery Point campus and Rip worked for a company named Sonalysts in nearby New London. In the summer, we moved out and let the owners use the house or rent it out for four times what we were paying. I would return each summer to our home in Florida, where all our personal belongings resided, and Rip would live in military housing at the nearby Navy base until time for him to do the work in Korea he did for Sonalysts each summer. For fifteen years we maintained this strange arrangement.

Among the several blocks of picturesque vacation homes, there were no schools, stores, gas stations, restaurants, libraries, or any of the other buildings usually found in towns large enough to have a name. We had only a police station, a fire station and a building called the Casino, which had nothing to do with gambling. The Casino featured one large empty room, used only on rare occasions for events such as wedding receptions or yacht club gatherings. A small part of the building opened up in the summer as a snack bar serving ice cream.

The tiny post office in the very front was open year around. Residents had to pick up mail from their post office boxes. We were residents of a sort – at least close enough to being residents to have a key to the post office box assigned to our address during the nine months of each year that we were strange renters.

Maybe I should more correctly refer to us as *strangers* who were renting. I say this because of an experience we had when first looking for a place to live in one of the nearby towns. Every house in near-by Noank bears a plaque displaying the name of the original owner and the year the house was built. Some residents there are so proud of the colonial history of the homes and the long tradition of passing down homes to people whose ancestors have lived in the area for generations that we horrified a person out working in his garden by inquiring about renting in the neighborhood. His immediate reply was, "We don't rent and we don't sell and those who do can go to hell!" Fortunately for us, his sentiment was not universal, although I did experience a similar resistance to newcomers in Groton Long Point in an encounter with another older resident.

One beautiful day in early fall, I had wandered out onto a jetty in our neighborhood that stretches out into Long Island Sound to sit alone on the rocks and write poems. After a while, an old man came along who asked me what I was doing there. I told him I was writing poems. He asked in a rather grumpy tone, "What kind of poems?" I read to him the poem I had just written about a rose bush that I had not recognized as a rose bush on first seeing it. He nodded, as if he approved the poem, and said, simply, "Okay," apparently having decided that I might actually be a good person to have in town. Then he just turned around and wandered off to continue his walk.

During our fifteen years of being strange off-season renters, I got to know the postman very well. He knew all about the translations and my efforts to learn Russian. One day he told me a woman who spoke Russian had moved into the neighborhood and told me approximately where she lived. I was eager to find her and to try to become friends.

My preferred form of exercise for decades has been bike riding. Riding around our quaint, un-commercialized neighborhood was a daily activity in good weather. On one of my rides, I spotted the woman I thought might be Russian. I stopped my bike and asked in my newly-learned Russian if she were Russian. I don't recall whether she just answered *yes* in Russian or tried to tell me she was actually Ukrainian, but we were on our way to becoming friends within minutes.

Anna came home with me that day and I bombarded her with my dreams of learning Russian and of translating the poems and plays. The picture of Jesus with the Russian words on the refrigerator cemented the friendship. She asked for a copy and when I visited her home sometime later, I saw that she had added it to the collection of images in her icon corner.

Anna agreed to help me with my Russian language efforts. She tried to help me with pronunciation by reading into cassette tapes poems by Anna Akhmatova. I had hopes that listening to the tapes might help me learn a more interesting vocabulary than I was finding in the travel tapes. Her voice was very soft, though, and I was often not able to hear clearly the words that she read.

Her friendship enriched my life through much more than just her help with learning Russian. The home Anna and her husband and two lovely daughters were renting was just a temporary home. They were having a larger home built not far away and we have wonderful memories of many visits there with them and their friends, celebrating the christening of their home and Christmas and Easter in the best of Ukrainian tradition. We danced folk dances in the living room, enjoyed a candle dance performance, toasted the arrival of Christmas hour after hour as it moved through different time zones with red wine that Anna was proud to show clung to the glass as it slipped down the sides and delicious, high-quality chocolates. At Easter, we played the traditional game of trying to tap uncooked decorated eggs without breaking them. There was always room for one more at her table. Whatever food had been prepared seemed to stretch like loaves and fishes to feed everyone who wanted to be there. I missed them very much when her husband was transferred to another state.

In the midst of my various efforts to learn Russian, Michael and I stayed busy working on the poem translations. We gave a talk on the poems and the translation process at the MAA/AMS Joint

Meetings in San Antonio in January of 1998. Happily, the poems were well-received. Perhaps more significant to me personally, however, was the fact that a publisher, Klaus Peters of A. K. Peters, Ltd., told me afterward that he had more interest in publishing my poems than the Kovalevskaya poems.

For some reason, a mathematician with a German accent had read a few of my math/physics poems at the podium before the talk. This was not an *Evening of Poetry* event. It was listed in the conference program only as a talk about the Kovalevskaya poems. I think the German professor and I must have been talking about my poems as people were gathering for the talk and she must have asked if she could share a few poems with the others.

In any case, I was certainly happy to hear the encouraging words expressed by such a well-known publisher. His words inspired me over the years to continue writing the kinds of poems he had heard that day. I have written more than a hundred math or physics-based poems and have been a presenter at many of the MAA/AMS Joint Meetings poetry events.[18] I have also written poems related to each of the sculptures in the book *Helaman Ferguson: Mathematics in Stone and Bronze*.[19] One poem defining a group was included in a *Scientific American* article that has been republished in several languages.[20] JoAnne Growney has included many of my poems in her extensive, growing collection of math-related poems: *Intersections – Poems with Mathematics*.[21] A French journalist, Pierre Carrée, in an article in *Des maths (mais pas seulement) pour mes élèves (et les autres)* published the poem *Point of Distinction* along with two of my symmetry-based artworks.[22] Several poems have been published in the *Humanistic Mathematics Journal* and *The AMATYC Review*.

Michael and I were invited to share the Kovalevskaya poetry translations in St. Petersburg, Russia in May of 2000. There, we presented the first ever English translations of her poems at an international symposium celebrating Kovalevskaya's life and her mathematical contributions.[23] The symposium was a gathering of mathematicians from many countries who presented papers on new mathematical results based on Sofia Kovalevskaya's work in differential equations. We were the only American participants and we were presenting poems – not high-level mathematics.

I attended talks throughout the symposium and noticed two things that differed from talks I had attended at MAA/AMS Joint Meetings. First, each presenter began with a long list of acknowledgments giving credit to mathematicians whose prior work had led to their result. Second, attendees listened with an ear for errors and it was expected that mathematicians who spotted errors would speak up. The revered mathematician, Olga Aleksandrovna Ladyzhenskaya,[24] was one of the persons I heard help a young mathematician by correcting a mistake he had made during his talk. I was told at some point that she was so revered that she generally had an entourage of young mathematicians traveling with her to whatever conference she decided to attend.

Ours was the best-attended talk of the conference. A few of the St. Petersburg mathematicians had even brought their daughters to hear the poems. I had prepared a few posters, using photographs of people in Sofia Kovalevskaya's life, to put on display before the talk. When I introduced the poems with a short biography, I was shown the same respect as other speakers when it came to noting any mistakes needing correction. I had used the expression *marriage of convenience* to describe the marital arrangement Sofia and Vladimir had agreed upon to help her further her education. Their arrangement, by mutual agreement, involved no physical marital relationship. The director of the Euler Institute, Ludvig Faddeev, corrected my mistake and said that it should be referred to as a *nominal* or *fictitious marriage*, which is very different from what would correctly be called a *marriage of convenience*. I was thankful for the correction.

That was actually the moment I felt most comfortable. Our abstract had stated that we hoped to receive feedback and suggestions from the attendees that might improve our translations. I was very happy to see that we could expect to receive suggestions, if our English translations were not true to the Russian versions. To my amazement, the translations received only praise. Our talk led to many encouraging conversations during the gatherings for meals and during a very special gathering for toasts to Sofia Kovalevskaya with red wine or vodka over plates of smoked salmon and caviar. Later, I enjoyed exchanges of e-mails with mathematicians who were interested in our project or who hoped I might be able to help them with translations to English of their writing projects. I began two friendships with our Russian hosts that continue to this day through e-mail, Christmas and birthday notes and Facebook messages.

Every moment of the trip to Saint Petersburg was part of an exciting and enlightening adventure – including the pre-symposium part that proved to be a bit rocky. Even as I am about to recount a few problems, I am completely mindful that our family might never have had the wonderful experience of traveling to Russia had Michael Dutko and I not been invited to present our translations at the *International Symposium Dedicated to the 150th Anniversary of the Birth of Sofia Kovalevskaya*. The symposium was very well organized and we appreciate all of the work that went into making it a success.

My younger daughter, Sarah, had just completed a bachelor's degree in history and, as a graduation present, we had decided to take her along on the trip. As always, we enjoyed her company, but it turned out that our special request for a family suite with two adjoining rooms at a hotel that had prepared only for a group of individual mathematicians complicated our stay. Our decision to let ticket prices guide which days to travel also added complications, since our flight arrived a couple of days before the planned start of the symposium. Initially, we were on our own, without the carefully planned assistance that had been prepared for the symposium participants, such as help with exchanging money, arranging for meals and use of public transportation.

Solely because of our special travel requests and our money-saving ticket arrangements, the two days before the other mathematicians arrived were pretty miserable – so much so that we actually thought of coming back early. Due to our special request, we were placed in the oldest part of the hotel. We learned later that other symposium participants who arrived after us had very nice single rooms in another section. Our room didn't even have hot water unless we let cold water run for about forty minutes. Our toilet, telephone, refrigerator and TV did not work at all. There was only a folding make-shift bed for Sarah to sleep on and no door between her space and ours.

I use the word *space*, because the two spaces were not large enough to be called rooms. The linens for her bed seemed to have been stored in a little free-standing closet for decades. The beautifully patterned old blanket was, unfortunately, full of dust. Rip and I had twin beds with cardboard headboards and similar blankets. It was very warm in the room, but a radiator heater was running that we could not turn off. So, we opened the window only to allow into our room a storm of mosquitoes and the sounds of a lively holiday celebration in a nearby parking lot – complete with very loud fireworks!

No one at the hotel spoke English and no one would take our traveler's checks, American money or credit cards for meals. All I had to eat the first day was a package of peanut butter crackers from home. But, because my Aunt Sara had sent me a book she had found at a used bookstore full of phrases the author thought a traveler to Russia might need, I was able to communicate the problems with the room in Russian well enough to get most of them fixed. We had laughed at the book when we had received it, thinking no one would ever need to say the things included, but we used many, many phrases from the book during those first two days. We also laughed when the man trying to help us with all the problems said the only word he knew that we recognized. He shouted "Hallelujah!" when he finally found a power converter that made my curling iron work!

When Elena, the Euler Institute organizer in charge of orchestrating the flow of activities for the symposium, saw our rooms, she was very upset. She could not understand why the hotel had given us rooms in such bad shape. The blankets seemed never to have been washed and the towels hanging on a vertical pole in the bathroom were what Americans would call tea towels and seemed to have been hung to dry by previous guests after use. When we visited Michael's room, we saw that his was not like ours at all. He had clean linens, fluffy white towels, new carpeting and a modern marble bathroom.

Elena went to the director of the institute and asked permission to move us to the apartment within the institute reserved for special guests. We had to wait until registration was complete to be sure none of the other participants actually met the description of important persons normally housed in those elegant rooms, but once the move was okayed, we walked our embarrassingly large amount of luggage to the wonderful accommodations inside the Euler Institute.

There we had two bedrooms, a lovely, large, wood-paneled bathroom with an efficient built-in water heater. There was a huge additional room with a large-screen television, on which, if I remember correctly, we were able to see American news channels. There was ample seating for relaxing, studying, or viewing the television. There was a large, well-equipped kitchen with a table that would be comfortable for hosting a dinner party for eight or more people. The floors were parquet and the windows, as I recall, were latticed like castle windows.

Except for the fact that the heat seemed not to have been turned on, we were beyond comfortable for the rest of our ten-day stay. We were even charged only half what the hotel had been charging. In addition, I was given a private office with a computer. It was in another part of the Euler Institute, but within easy indoor walking distance.

PRIVATE OFFICE AT THE EULER INSTITUTE

I think every participant was given access to an office and a computer. Mine was used mostly for keeping in touch with friends and family back home by e-mail, but that was an important part of feeling comfortable during our stay. Once we were settled into the new place, we enjoyed our visit very much. We went with the group to a chandeliered concert hall where we heard Tchaikovsky performed and to a ballet performance. We enjoyed hours at the Hermitage Museum, while wearing our blue plastic shoe covers to protect the beautiful parquet floors. We visited the Winter Palace with the group and the Summer Palace with a friend we had made among the participants. We saw two of Sofia Kovalevskaya's homes from the outside during a bus tour of the city.

KOVALEVSKAYA HOME IN SAINT PETERSBURG
PHOTO BY SANDRA D COLEMAN

The three of us were treated to private visits to several monuments, a very unique restaurant and to the main downtown shopping area by one of the St. Petersburg mathematicians with whom we are still in touch. Each participant had been assigned a personal guide to help with navigating transportation and sightseeing in the city. The thoughtfulness of our hosts in arranging for this was very impressive and very much appreciated.

We were welcomed warmly from the first day of the symposium and began to make friends with mathematicians from all the different countries represented right away. Kjell-Owe Widman, the head of the Institute Mittag-Leffler in Stockholm, reached out to me and said that I could reside there and do research any summer I liked. He, Rip and Sarah had stayed out late enjoying each other's company on several nights, while I quieted down for the night and caught up on e-mails in my little private office. The Institute Mittag-Leffler, the former home of Gösta Mittag-Leffler, is now a place where researchers can find original documents related to

Sofia Kovalevskaya and even such unique treasures as a tablecloth she embroidered.[25]

Elena also said that I would be welcome to come and stay at the Euler Institute at any time to continue my research on Sofia Kovalevskaya's life and writings. I received a formal invitation to speak at another Russian conference in the Ural Mountain region, but I was not sure I had the courage to try to make that trip, since the three charming mathematicians inviting me spoke very little English and I was not confident I would be able to get help if any problems were to arise such as those we had encountered during the first couple of days in St. Petersburg.

Because I had been talking to the participants all week about the translations and my efforts to teach myself Russian, I viewed the audience for our talk on the last day as a group of friends eager to hear more of my stories. I was very calm and comfortable – even more so than at the MAA/AMS conference in San Antonio.

I had tried speaking a bit of Russian now and then during our stay and had impressed a few people with renditions of poems I had worked hard to memorize. A newfound friend gifted me with her personal copy of Anna Akhmatova poems in support of my idea that I might learn poetic vocabulary better from studying poems than from listening to taped lessons meant for travelers and business people. I smile now when I think back at being cautioned not to say anything in my funny Russian accent in public places like museums. Apparently, if I had opened my mouth, our entire group would have missed enjoying some benefits of the experiences we were being treated to, since tourists are treated differently than locals. I certainly did not want to stand out as a tourist, so I was obedient and did not speak any "funny" Russian outside the Euler Institute!

I came home full of enthusiasm for Kovalevskaya research and was hopeful that Michael had also been inspired. I stopped hearing from him, though, until he called one day and asked us to join him in Groton for dinner. He met my husband and me at his favorite restaurant and we enjoyed a nice meal, but when we said good-bye at the end, I had a feeling his hug was the kind of hug you give when you may not get another. He had shared with me some health concerns earlier, but without much detail. When I did not hear from him for a long time afterward, I began to worry a lot about what was wrong. I looked to see if he was listed in the schedule of classes at Scranton for the upcoming semester. He was not. I think I knew before looking further that he was gone. I can't recall all the steps I went through to try to find out what had happened, but I eventually learned that he had died not long after that dinner in Groton. He must have known then, but didn't tell me. It was not easy to go on with translating without him, but eventually I began again. I was totally on my own with the next project I had in mind.

It had been the mention of her poems in many accounts of Sofia Kovalevskaya's life that had led me to want to complete the poetry translations with Michael. A second piece of literature almost always included in any discussion of her writings was a pair of companion plays titled together as *The Struggle for Happiness*. The first of the two plays describes *How It Was* and the second *How It Might Have Been* in the lives of six young people faced with important choices at critical moments in their lives. Summaries stated that the characters were essentially the same in each play, but with enough difference in their views on life and love to cause them to make significantly different choices at critical moments. The brief descriptions had made me want to read them, but, again, there were no English translations, so I decided to try to translate them myself, although I knew the effort would be very difficult without Michael's help and would take a very long time.

Initially, I wanted only to know more details about the content of the plays. At first, I had planned to translate only for my own satisfaction and began working on a rough translation of the first play long before the translation tools now so readily available to translators were available to be put in my toolbox. Once more, I relied on structure to locate the dialog pages of *How It Was* among the hundreds of pages of Russian in my *VP* gift box. I purchased and used an early version OCR scanner to try to transfer the words on the pages of the play to my computer, but, unfortunately, because of the many physical flaws in the paper copies, the resulting digital pages were chock full of spelling errors and distorted formatting.

To correct the errors, I had to learn to type in Russian. I installed the Russian keyboard available in Microsoft Word, which allowed me to switch back and forth between Russian and English. I had no Russian keyboard overlay, so I memorized the locations of all the

Russian letters. After many weeks of working on comparing the scanned pages to the copied pages and making hundreds of necessary corrections, the digital images on my computer finally matched the paper pages from which I was working.[26] From there, I was able to begin the slow process of translating the first play.

For months I vacillated between periods of enthusiasm and discouragement. There were times when I let so much time pass without working on the plays that I had trouble remembering which of my growing collection of computer files bearing similar names held the most current work. Sometimes it was only eagerness to know what would happen next in the play that brought me back to the translation project. My Russian was not good enough to scan the dialog quickly, so to know what would happen next, I had to translate.[27]

Once, I sought out the help of a very kind math professor at the Avery Point campus who spoke Russian. By the time we met, I had memorized a couple of Russian poems by Pushkin and Lermontov. I recited them for him one day and he told me my accent was more Ukrainian or Polish sounding than Russian – softer and less harsh. I suppose I sounded like Anna. He, too, was Ukrainian. He spoke Russian, but seemed displeased at having been made to learn Russian in school.

By the time I sought his help, I had made enough progress on the first play to be wanting feedback on the accuracy of what I had written and I needed help with a few troublesome passages. I asked him if he would be willing to proof what I had written and make suggestions. He was happy to help, but he was very shy about talking face-to-face about suggestions. Instead, he used the editing software in Microsoft Word. He worked on pages I had given him for several days and nights and made many notes in the margins suggesting changes. His sincere effort to be helpful happened just before one of my summer trips to Florida. I was reluctant to make any of the changes without actually discussing why one phrase was better than another. I was so frustrated over having to leave town without getting to discuss his suggestions that I was stifled rather than helped in my translation efforts and ended up stopping work for the whole summer.

In the fall, I returned to my work on the play. I felt triumphant when I finally completed the first rough translation of *How It Was*. I had reached a major goal in just being able to know the content. Based on the title of *How It Might Have Been* and summaries I had read, I thought I knew how the next plot would play out, so I rested for a while before beginning on what was likely to be a very long second effort.

When at last I did start on the second play, I discovered that the plot was very different from what I had imagined. The summaries I had read early on had actually been a bit misleading. Once I became engaged in the second plot, translation became an exciting activity again. I began to wonder once more what would happen next in the story. Still, the process I was using to translate was very, very slow.

Translating a play slowly is nothing like just reading a play in your own language. The only experience I can compare it to is the experience of exchanging letters with a blind person who used to type long letters to me in Braille. I could not possibly scan the pages of his letters quickly and had to decode the words one at a time, which I did more quickly than usual the one time he was relating to me details of a romantic encounter he had had.

As the characters and love-story plots unfolded in the plays, I found both *How It Was* and *How It Might Have Been* to be more exciting than I had expected them to be. I was eager to follow the stories. I was also discovering that, within the two plots, Kovalevskaya was unfolding a story involving complicated romantic and family issues in the context of practical problems encountered in the earliest days of the industrial revolution. She was intentionally addressing political differences in opinions on how workers should be treated, other social issues involved in the ongoing struggles to transcend class and gender boundaries, and a number of cultural issues involved in the ways religious beliefs and traditions affected reactions of people to new social and scientific ideas. These aspects of the plays add to the history of late 19th century Europe. I was able to imagine how parallels between the issues in the plays and issues of today could make the plays useful tools in classrooms for initiating dialog on a broad range of current topics. At some point between beginning work on the first play and completing the second, I decided the plays were worth translating in a more polished form – one more suitable for sharing and possibly even for performance.

Through the years of working on the translations with primitive tools, I was encouraged by every form of new technology being developed that might be helpful to translators. A tool called ScanMarker allowed me to speed up the transfer of the pages of the second play to my computer. Using that device was a huge improvement over the process I had used on the first play with the OCR scanner – the process that had presented me with the need for hours and hours of tedious typing as I corrected errors on the scanned pages. Obtaining a digital copy of the pages by whatever means was an essential first step before I could use any sort of translation tool to decipher the words, whether a dictionary or any of the various translation software tools beginning to be available.

It was not until I was nearly finished with the translations that the resource I had needed so much became available to me. My husband, always on the lookout for ways to help me, had persevered until, at last, he found the full text of *Vospominaniya Povesti* on-line. It may have been just released to the public. By the time we found the digital text, I had retired from teaching and we had moved permanently to our Florida home. Having access to the *VP* text in a form from which I could copy and paste to translation software, and being free from the duties of teaching, I was able to accomplish more in a few months than I had accomplished in the previous many years of work.

Most of my time after that point was focused on refining the dialog to be sure there were no significant errors with regard to meaning in the translations. I had heard jokes about the dangers of translation software interpreting a passage like "The spirit is willing, but the flesh is weak." as "The wine is good, but the meat's awful." After twenty years of research, I felt very sure that I knew Sofia Kovalevskaya well enough to convey the words and messages she intended without a troubling number of nonsense errors.

Eventually, I felt confident that the translations were close to being ready, but I was still reluctant to share them publicly, because I felt there was one thing lacking. There was more work to be done to make the plays accessible to a modern audience. The austere formality of language, normal for 19th century conversation, was obscuring the actual closeness of the relationships among the young people portrayed. I felt that modern audiences would not be likely to become personally engaged with the plot or the characters unless the language could sound less formal.

I made the bold decision to modernize the dialog very slightly, making the translations more accessible to modern-day audiences. I was careful to retain the variations in ways of speaking that reflected differences in levels of formality among characters of different generations and different positions in society at that time, but allowed contractions, for example, in informal dialog among the six main characters to make it easier for audiences to relate to the younger characters. Just such small changes made the plays more readable and much more likely to be performed. I felt strongly that Sofia Kovalevskaya and her co-author would be pleased. They would likely want to present these plays as much today as they did over one hundred years ago.[28]

Kovalevskaya and her friend, Anna Charlotte Leffler, wrote the plays together during the period when Kovalevskaya was teaching at the University of Stockholm in Sweden. Anna Charlotte was the sister of Gösta Mittag-Leffler. He had also studied under Karl Weierstrass and was well aware of Weierstrass' high regard for Kovalevskaya's mathematical ability. His efforts on her behalf were instrumental in securing for her a Stockholm University faculty position – the first such position ever offered to a woman mathematician. He also opened to her the position of editor for the renowned journal, *Acta Mathematica*.

Anna Charlotte was already a well-known writer when Sofie asked her to do the actual writing of the Swedish versions of the plays that she had been creating in her mind. Sofie described to Anna Charlotte the plots, characters, and dialog. Kovalevskaya definitely sees herself as the creator of the dramas. In fact, she is said to have first realized that she had talent for writing as these plays unfolded. *A Russian Childhood*, her much-acclaimed memoir, and the semi-autobiographical novella, *A Nihilist Girl*, were written later.

Sofie and Anna Charlotte were very close during the time they were writing the two plays. Both were extremely excited about their collaboration and they worked quickly to produce the first drafts.[29] Completing the plays was so important to Kovalevskaya that she took a break from her work on the Prix Bordin solid body problem to focus on their writing.

Her efforts on both the plays and the Prix Bordin problem were further interrupted by the health problems and political complications occurring in her sister's life. Aniuta was very ill and Sofie took a leave of absence to be at her dying sister's bedside in Russia. In a letter Sofie wrote to Anna Charlotte from Russia, she said that, during periods when her sister was unconscious, she found escape from her anxiety by reading Poincaré's mathematical papers.[30] Ideas for the play developed at Aniuta's bedside and it seems those ideas were influenced by a principle discovered in Poincaré's research on the three-body problem.

Kovalevskaya and Poincaré were developing ideas in their analyses of dynamic systems that are referenced in Kovalevskaya's preface to the two parallel dramas. She likens the effect of the two Karls' slightly different personalities in the two dramas to the sort of difference in initial conditions that can significantly affect outcomes in mechanical systems, such as the trajectory of a pendulum or the behavior of the third element in the three-body problem.

Kovalevskaya was reading about Poincaré's discovery of the importance of *extreme sensitivity to initial conditions*. Poincaré proved that determinism does not ensure predictability. His discovery of unpredictable behavior in deterministic systems formed the basis for the development of modern-day chaos theory. In his biography of Gösta-Mittag Leffler, Arild Stubhaug[31] explains, very clearly, how Poincaré first discovered chaotic behavior in celestial mechanics as part of his prize-winning research on the three-body problem. This was a discovery reflecting the sort of creativity and imagination Weierstrass admired.

Weierstrass considered imagination and creativity to be traits essential to a great mathematician. In the letter to Kovalevskaya[32] that includes his famous statement that *one cannot be a great mathematician without having the soul of a poet*, he is saying that he sees mathematicians as falling into one of two categories — those who have something of the soul of a poet and those who do not. The poetic traits he considered essential for greatness, he saw in both Kovalevskaya and Poincaré. It is, at least in part, the depth and breadth of Kovalevskaya's mathematical insight that influenced his view of her as a great mathematician — a mathematician with the soul of a poet.

The translations of the plays and poems are a small part of an on-going search for the soul of a poet in Sofia Kovalevskaya. The plays make her views on social and political movements of her day very apparent. They show great sensitivity to the kinds of hard choices people often face in seeking a satisfying life. They may even be somewhat auto-biographical, in that the character, Alisa, expresses in dialog personal feelings Anna Charlotte claims Sofia expressed to her with regard to her own struggle for happiness and the struggles for happiness of her sister, Anna Jaclard.

There is still much to be discovered in the hundreds of pages of Sofia Kovalevskaya's writings in *Vospominaniya Povesti*. For now, I hope the translations of the plays and poems, along with the commentary, will provide engaging reading and, perhaps, some new biographical and historical information.

It is exciting to imagine that Sofia Kovalevskaya's success in both mathematics and writing could inspire young women working on developing their interests and talents in mathematics to explore and develop their own other interests in such a way that people may, once again, say that they are surprised that a mathematician can excel in two fields at once.

September 2021

THE POETRY OF SOFIA KOVALEVSKAYA

Seven poems written by Sofia Kovalevskaya were published in Russia by her fiancé, Maxim Maximovitch Kovalevsky,[33] shortly after her death at age 41. He gathered together and published all of her literary works known to exist at the time. This huge volume of writings was revised and edited by P. Ia. Polubarinova-Kochina as *Vospominaniya Povesti (VP)* in 1974.[34] The collection now contains nine of Kovalevskaya's poems, two of which were found among her daughter's papers after the daughter's death in 1952.[35]

Kovalevskaya is known to have written other poems. She talks about some of them in the story of her childhood. In *A Russian Childhood,* she wrote that at one point she "was unshakably convinced that she was going to become a poet," saying:

> *I loved poetry with a passion. Its very form, its very rhythm delighted me. I greedily devoured every excerpt from Russian poets that caught my eye and, I have to confess, the more high-flown the poetry the better I liked it...The very beat of poetry enchanted me so much that I began composing verses myself at the age of five. But my governess did not approve of this occupation. In her mind was formulated a perfectly crystallized concept of [the] normal, healthy child...and poetry-writing did not accord with this concept at all.* [36]

Kovalevskaya went on to say that in spite of the fact that the governess mercilessly hunted down all her poetic efforts, pinning the scraps of paper to her shoulder and declaiming each piece in a loud distortion before her brother and sister, she was not in the least discouraged in her dream. She continued to compose poems and to recite the poems to herself until she had memorized them, since her dread of the governess' reprisals made her reluctant to write them down. Such a pity! It would have been interesting to read *The Bedouin's Salutation to his Horse* and *The Sensations of a Pearl Diver* and the finished first ten verses of the hundred-twenty-verse epic she had planned to write. Those early poems have probably been lost forever, but *VP* does include the nine poems that have been preserved in writing. Those are the nine included here.

In the nine months Michael Dutko and I worked together to translate eight of the nine Kovalevskaya poems, our primary aim was to add to the personal image of Sofia Kovalevskaya. For this reason, although it is neither unusual nor unacceptable in poetry translations for the translator to convey the poet's thoughts in a totally new form, we decided that it was important for us to adhere as closely as possible to Kovalevskaya's original forms. Always being careful not to sacrifice content or meaning in the effort to match rhyme scheme, meter and other poetic elements, we worked together to develop translations that sound very much like the Russian versions when read aloud. The process involved a succession of first attempts followed by many revisions until we felt we had reached our goal. By adhering to both form and content, we hoped readers would be able to decide for themselves what the poems reveal about the poet.

Chameleon was the first of the Kovalevskaya poems I translated. The translation was completed before I met Michael and is based on a free-verse Russian version of the poem found in *VP*. The poem was originally written by Kovalevskaya in Swedish in response to a rhymed birthday greeting given to her by her friend, Anna Charlotte Leffler. The initial version of *Chameleon* was rhymed.[37] Both rhymed and unrhymed versions include a teasing reference to Anna Charlotte's attempt to employ rhyme in her birthday message. In *Chameleon,* Kovalevskaya portrays herself as someone as changeable as that animal, a woman fully capable of adjusting to the various needs and expectations of many different friends, but only able to glow when surrounded by inspiring, encouraging companions. In *Love and Mathematics*, Kochina shares a quote from Maria Jankowska's recollections of her friend, Sofia Kovalevskaya, who remarked that this chameleon-like quality was, indeed, a part of Sofia's nature.

> *Each of Sofia's numerous friends preserved in his or her memory a different image, because to each she presented herself in a completely different light. But there was not the least falseness in this; it was just that her rich nature gave to the person who interested her at that moment exactly what seemed to her to suit that person.* [38]

Chameleon

You've known the chameleon since childhood.
When he sits all alone in his corner,
he seems so plain, so unattractive,
gray. But in good lighting he
can even be beautiful.
He doesn't possess his own beauty; he just
reflects all that he sees around him
that is beautiful and good...
He can shimmer from gold
to green to blue. As his friends are,
so, will he be.
In this animal, it seems,
I see my own likeness. My dear
friend, wherever you lead, I
will always be right behind you;
I will not fall behind; I will never give up.
If one has a friend like you,
one has to pay for that honor.
You write, you paint, et cetera!
For me, this is child's play, *patatras*!*
But, Lord have mercy! Now you want
to try to write poetry!

* a snap! (French)

Some of Kovalevskaya's poems are highly structured with especially strong use of rhyme and meter. In two of the poems, *A Husband's Complaint* and *A Humorous Message to V. O. Kovalevsky*, rhyme and lyrical rhythms seem to aim at creating a light-hearted tone with the intent of softening her serious messages of complaint. In these two poems, both of which were intended for her husband, her aim was clearly to convey her complaints in a form she thought might get a better response due to the humor she hoped he would find in seeing her thoughts all forced into rhyme.[39]

A Husband's Complaint is written in the voice of her husband, Vladimir Kovalevsky,[40] who, from the picture painted in the poem, seems to have reneged on the initial agreement to be content with a fictional marriage and is still not happy even after having succeeded in changing the arrangement.

Their marriage had supposedly been entered into for the sake of furthering Sofia Kovalevskaya's educational opportunities. The larger plan was that the marriage would also provide a way for other women in Sofia's circle to travel as her companions to countries accepting women students at universities – thus achieving a goal she and Vladimir shared, the goal of making progress toward improving society through education for all. Vladimir had agreed to the fictitious nature of the marriage, meaning there would be no expectation of a physical relationship.

In *A Convergence of Lives*, Koblitz sites a number of opinions with regard to Vladimir's true feelings for Sofia.[41] There have been various speculations on the level of importance his support of her educational aspirations played in influencing his desire to choose her, rather than one of the other two women in the group of three who asked him to marry any one of them so that all could obtain passports enabling them to travel and study abroad.[42]

Some accounts present Vladimir as being fully aligned with the goals of the young idealists who were willing to sacrifice personal satisfaction for the sake of the noble cause of creating greater educational opportunity for all. But in *Love and Mathematics*, Kochina shares a letter[43] written to his brother soon after the agreed upon fictitious marriage that would indicate that his admiration, possibly love, for Sofia was overwhelming from the start.

> *I cannot comprehend political and economic problems half as rapidly as she does…I think that she will make me a decent person…I cannot conceal from myself that her nature is a thousand times better, more intelligent and talented than mine, to say nothing of her diligence…On the whole, she is a young phenomenon.*

Kochina also notes that at about that time Sofia wrote to Aniuta,

> *Brother is very dear, nice, and good, and I am singularly attached to him…You won't believe how he cares about me, courts me, and is ready to submit his every habit and desire to mine…I love him truly with all my heart, but as I would a younger brother.*

In the poem, *A Husband's Complaint*, Sofie imagines Vladimir speaking as a discontented husband, who has found himself wanting more in spite of the fact that he had agreed to a nominal marriage. She paints an image of herself as an irresistibly beautiful young charmer, whose husband tricked her into marriage with false promises, knowing he never planned to adhere to his vows. She also has him viewing her as an intelligent woman, who fully expects her husband to abide by any promises made – a woman whom her husband would do well to treat with care – and a woman whom her husband would not dare try to deceive.

A Husband's Complaint

I am fed up, I can tell you,
With complaints and moans, no end –
Women, saying fate and laws are
Always so unjust to them!
No! Believe me that the victims
Are not they, but we, the men,
Who much grief and sorrow suffer
For our families and for them!
How unwieldy is our burden.
Every man agrees it's true,
But I hope that this, my story,
Will make clear this truth to you.
Now, my youth I spent not dully,
As, I say, befits a man!
As were most of you, I'll warrant,
I was tempted now and then.
Beauty, wine, and love I savored,
'til I reached the point in life,
When I finally acknowledged
It was time to take a wife.
My misfortune was to meet a
Maid – a charmer from the start.
Such a wonder of creation!
Young and beautiful and smart!
I say no one could resist this,
And I fell in love, of course.
But, though she was very sweet, there
Was one reason for remorse.
Her sweet spirit was infected!

Various books had filled her head,
So that, rather than give in to
Me, she put me off instead,
Saying stubbornly to me, "I
Still am young. I've much to learn.
You can wait until I'm done with
That and then you can return.
Then, an equal with my husband,
I'll be confident and bolder."
I, in answer to this speech of hers,
Could merely shrug my shoulders.
For her little eye so sparkled
And the little rogue so charmed,
There was no way to refuse her,
For my heart she had disarmed!
I saw then: A ruse is needed.
This requires both skill and guile.
I'd approach this maidenly mindset
With an offer and a smile.
I convinced her I could help her
In her studies, as a friend.
For such innocent deception,
Who would blame me in the end?
…I repeated my entreaties
With assurances profuse,
Until, caught up in the rapture,
She could find no more excuse.
At long last, to me she yielded
And my wish was satisfied,
As this lovely maid consented
To become my lawful bride.
But in vain I hoped to gain a
A bit of peacefulness that way.
Before long, for my mistakes a
Bitter price I'd have to pay.
It turned out that the poor woman,
Who had chattered about rights,
Thought I'd meant each word I'd uttered
Right up to the wedding night.
And, naïvely, she's expecting
From her husband, even now,
That each word he said in passion
He should honor as a vow.

Oh! My friends! I tell you truly
I was frightened at the sound
Of a woman who'd so freely
On her thoughts on this expound.
There's no doubt a learned wife is
A calamity, I fear.
Husband! Don't wait for respect from
Her, and make your answers clear!
Share her doubts! Seek not to tell her
To be gone, and as to "business,"
Be frank about what's going on,
For she'll ask you with persistence.
As now anxiously I ponder,
I'm becoming more aware
Of the need to treat a woman
Who's intelligent with care!

The fictional nature of the marriage continued for five years. The reference in the poem to her becoming his *lawful bride* may be a reference to the point when the decision was made to begin living as man and wife during a period when Sofia was feeling very sad and lonely after her father's death. The lines about business indicate the poem may have been written during the financially troubled times when Vladimir was becoming involved in dishonest business ventures, which he tried to keep hidden from Sofia.

The next poem, *A Humorous Message to V. O. Kovalevsky,* was clearly written after the eventual consummation of their marriage and the birth of their daughter, Fufa. Here, Kovalevskaya describes herself as a dusky wench[44] impatiently awaiting her husband's return to her family's Palibino estate – or at least some response to the letters she has written from there. Though surrounded by family members: her sister, Aniuta, her sister's husband, Victor Jaclard, their young son, Yuri, her young daughter, who she calls Fifi in this poem, her brother, Fedya, and her dear friend, Julia, she is bored and is neglecting her mathematics to write rhymes that she hopes will get a response from Vladimir. The poem was written during a period when Kovalevskaya had returned to her Palibino home, seeking some much-needed rest and relaxation. Most of her family had gathered there, but Vladimir was apparently being annoyingly slow to join the group.

A Humorous Message to V. O. Kovalevsky

My friend! For two whole weeks now, every hour I've suffered pain,
As I've waited for your coming – for I've waited all in vain.
To my volumes of vexation, no response from you I've heard.
You have neither come nor written me. No not a single word!

It bores me so to have to wait and bear ill will for hours.
So, I thought I'd see if poetry might somehow have the power
To make you feel ashamed, you villain, for ashamed, yes, you should be.
You ought to have more feeling for our marriage and for me!

You must see this demon holds me in its claws. It's an obsession!
My muse will not release my captive soul from her possession.

Having laid aside all interest in my cookbook, integration,
My master's thesis work and Korkin's differentiation.[45]
I look for ways to vent my rage – all day I spin out rhymes,
And every hour to Parnassus toss my soul ten times.

Dusty notebooks full of poems Julia came upon one day,
Which long ago in some dark corner had been laid away,
And by the poems' appearance there, I was reminded of
A world of long-forgotten dreams and long-forgotten love.

And, so, it seems the muse has been awakened in my bosom,
So that I am ready now to trade for her Minerva's wisdom.
Yet, I admit that, though it's old, the proverb's true, so far
As it would say of poets: *qui a rimé – rimera.*

But, so far, this is prologue. No more hesitating!
My point in writing is to say how bored I am with waiting.

Here, it's quiet; there's no fighting. It's a calm and peaceful life.
No one complains and every day the colonel and his wife[46]
Just memorize as many foreign phrases as they can,
But this language is a thing, I fear, he'll never understand.

Yuri's better than his Papa – not quite reason to be pleased –
But that he's great, I'd have to say, most everyone agrees.
Our marquise, Tante Fifi, I will tell you, should you care,
At least two times a day, now, wants to change the sash she wears.

The youthful mathematician ponders Herbert[47] by the hour –
And, while he's thinking, bowls of cherries, blissfully devours.

We tried to tutor Julia in the art of playing cards,
But she has so little talent that it's going to be hard.
She has mastered being quiet – doesn't give away her hand –
But even Aniuta gives her trouble now and then.

Your dusky wench is bored with always waiting, I must say,
And goes running to the road outside at least ten times a day.

Every dog's bark, every jingle of the carriage bells she hears
Makes her tremble with excitement that her husband may be near.
She runs out in expectation, but then, being fooled again,
She just curses love and marriage and all mean, disloyal men!

From the reference to Fufa as a toddler and what is known about Vladimir's business troubles, we can guess that Vladimir was already deeply in debt due to his unwise business associations. He was probably avoiding having to face his growing financial problems and was likely avoiding having to discuss them with his wife or anyone else in the family. More details of the financial failures and mental state that led him to eventually take his own life in 1883 can be found in Koblitz's biography.

The next two poems, *The Unknown Singer* and *Did It Happen...*, should be considered a pair, since together they reveal something interesting about Kovalevskaya's approach to rhymed poetry. In much the same way she enjoyed developing proofs and putting them into elegant form, she may have enjoyed the challenge of expressing her thoughts in two different ways – first in prose and then in rhyme, as though she were playing a game with words. The first of these two poems tells, in a form she calls poetry in prose, essentially the same story the second poem tells in metered rhyme. Both express with great passion the pain she suffered because of an unfulfilled relationship.

The man to whom these poems allude was most certainly not her husband, but was more likely Maxim Kovalevsky, a distant cousin of her husband and someone to whom Kovalevskaya felt an immediate and very strong emotional attachment on meeting him after her husband's death.[48] She compares her frustration at not being able to complete what was begun in this relationship to the annoying feelings that come when a person tries in vain to recall long-forgotten words to a strangely familiar song. She paints herself as being preoccupied with this relationship in much the same way she might have been preoccupied with a challenging mathematical problem. Much of the pain she experienced with him may have been due to her reluctance to let an intriguing problem go unsolved.

Their relationship was complicated and was often a source of emotional stress for Kovalevskaya, but, it seems that, during a visit Sofia made to Maxim's villa in Beaulieu-sur-Mer on the southern coast of France in December of 1890, they had made the decision to marry and were aiming for the following spring or early summer. It was on the way home from that visit that she carried her own luggage in a cold rain at a train station, because she didn't have the local currency she needed to pay a porter.[49] She was very ill by the time she returned to Stockholm and died shortly after the trip. Maxim confirmed in letters that they had planned to marry and made a sincere effort to be allowed to adopt young Fufa.[50]

I traveled to Beaulieu-sur-Mer and was able to locate Maxim's villa overlooking the French Riviera very easily by just taking a short walk up a beautiful hill, fragrant with orange blossoms. There is a plaque on the wall by the door identifying the villa as the home of Maxim Kovalevsky. It includes a reference to the writer, Chekov, as a frequent visitor. The home is now used as a school, but seems to have been given to the Red Cross at some point. Fufa never married, but she worked with the Red Cross, so there is probably a story there about how the property changed hands through the years. The beautiful view Sofie and Maxim would have had is now blocked by an apartment building. I was able to speak enough French to persuade a resident there to give me access to a place from which I was able to photograph Maxim's villa. The school was closed on the day I was there, but I have dreams of returning someday and being allowed to go inside.

MAXIM KOVALEVSKY'S VILLA   PHOTOS BY SANDRA D COLEMAN

The Unknown Singer

The unknown singer fell suddenly silent, and the song which had had no beginning was now also without any end.

Oh! How unbearable was the silence surrounding me at that moment! It seemed that the strings of my soul, playing in unison with the music, had suddenly snapped!

I remember how for days, thereafter, the barely discernible melody continued to follow me, independent of my will, sounding incessantly in my ears.

Sometimes, I was able to reconstruct single words, excerpts of phrases, ends of melodies. It would seem to me that with one more effort the entire song might restore itself to my memory. But in the next instant everything would become muddled again, so that the hidden meaning of this elusive song would remain an enigma to me forever – the key to it utterly lost.

Won't such be the fate of our encounter, oh, my fleeting friend? Won't our meeting be like a flash of lightning, illuminating me just for a moment, only to be extinguished almost immediately, leaving me once more in a dungeon of darkness?

It was, after all, a simple chance event that brought us two together and another such event which later parted us again. How formal and how cold was our farewell!

There was, no doubt, a moment when we felt ourselves to be so close, so much alike, so essential to each other. But not a word was spoken, as though we two became mute at that moment.

And so why, although not one word has been said, not one tear has been spilled, and no visible, tangible bond exists between us, why does it seem to me to be so unnatural that we are now parted?

I will, probably, get used to your absence. At first, your image will haunt and torment me often – like a lingering unsolved problem. But gradually it will become more and more vague; eventually, it will be erased completely, and my heart will yield submissively to the yoke of the cold and utter indifference, which at last it's prepared to accept.

Did It Happen...

I wonder, has it ever happened
As you walked aimless, unconcerned
Among the crowd, that some impassioned
Song, by chance, you overheard?

> O'er you waves gusted, unexpected –
> The memories of years gone by –
> An echo in your soul reflecting
> Sweet and intimate reply.

These are sounds you've heard since childhood –
Heard them time and time again –
Such contentment, joy, and torment
You remembered now in them.

> You hastened with your ear to follow
> Faint, familiar sounds you heard.
> Every single sound you longed for –
> Longed to capture every word.

Then, suddenly, the music ended.
The voice, without a trace, was gone.
Without ending or beginning
Evermore remained the song.

> Oh! How hateful seemed that moment –
> Silence covering everything –
> As though lack of sound had broken
> In your soul an echoing string.

Then, how tiresomely it plagued you,
Always following you around,
As though your ears would disobey you
And recall once more the sound.

> Sometimes coming out of nowhere

Bits of phrases, meter, words,

It seemed you might in one more moment

Grasp the whole of what you'd heard.

But, then, as soon, the song was over,

Just enigma in its place.

All you thought you'd come to know

Had disappeared without a trace…

Shall such befall our fateful meeting?

Will it, too, fade without a trace?

Shall that which flashed so bright now flee –

To some deep, hateful, dark abyss?

Simple chance brought us together

And may part us. Who can tell

If, now, chance the bond will sever

With indifferent, cold farewells!

It seems we, once, in one another,

Heard some dear, familiar sound,

But not a single word we uttered,

As if chains our tongues had bound.

So, without tears, without dejection,

We accept our separate lives,

Although sometimes from oblivion

I may see your shadow rise.

Hazy images will haunt

As shades of you return to me,

To trouble, worry, and to taunt

Like some, yet unsolved, mystery,

Until, in time, I no more see

Cherished features I adored,

And a more submissive me

Yields to cold, eternal void.

In the poem *Grunya*, Kovalevskaya tells the story of a young girl who, dreaming of being a martyr, imagines herself being led to a fiery death in the public square. A character in Kovalevskaya's one completed novella, *Nihilist Girl*, has thoughts closely related to Grunya's, although Grunya is a peasant girl and Vera Barantsov is the youngest of three daughters in an aristocratic family. Both characters had lost themselves in reading passionate accounts of the lives of Christian martyrs and both embraced the goal of earning a martyr's crown by being burnt at the stake. [51]

Grunya is another poem written first in prose and then in very complicated rhyme. In places, words intended to go together as rhyming pairs display the kind of hidden sound relationship heard in words like *recent* and *repent*. It was the effort to incorporate this rhyming element into the translation that led to the long delay in developing a satisfying translation of the rhymed version of *Grunya*.

Grunya

Having read about the lives of the holy martyrs, Grunya is constantly occupied with one thought: How to imitate them. In recurring dreams, she envisions herself in a scene of torture and execution. The images follow her everywhere.

In the rays of the blazing sunset, she sees the crimson glow of a fire. A large square is teeming with people. The air is filled with the deafening peal of a bell.

Grunya, in mourning clothes, is being led to execution. Fearlessly, cross in hand, she ascends the scaffold, but before her death the Holy Spirit speaks through her mouth.

All the people are troubled and amazed, but a scorching flame engulfs her instantly and angels carry her soul to heaven. In heaven, there is much jubilation...

The unrhymed version appears to be unfinished. The rhymed version, as it appears in Russian in *VP,* is included in the Appendix, accompanied by an outlay of the extremely complicated, somewhat irregular rhyme scheme that was so difficult to match.

Grunya

A saint's crown of suffering is being prepared
and her heart pounds within at the notion that, maybe,
an ending like this is a fate she might share.
The thought of this haunts her and gnaws at her daily:
that a crown everlasting, the crown of a saint,
she might, one day, acquire. In passionate daydreams
she sees herself often in torture and anguish.
Each day, in the fiery red rays of the sunset,
she sees a great bonfire, crimson and gleaming.
She hears the bells clanging and sounds of chains clinking,
and weapons resounding.
People file to the village and watch from the square,
as she's led to her death, calm, peaceful and clear.
She ascends to the bonfire in dark, somber clothing –
a cross in her hands – without cursing or moaning.
God's spirit provides her invisible strength.
Then, through her own mouth, the Lord can be heard
in the words Grunya speaks at the moment of death.
Troubled and stunned by the powerful words,
the people en masse start to weep and repent.
They tremble, prepared to believe what she spoke,
but the flames of the bonfire and bright, crimson smoke
engulf and surround her. The pain lasts a moment.
It's over and done and at once in the heavens
bright angels arrive to receive the pure soul.
She's surrounded by light, by angelic expressions,
as her heavenly home would her spirit enfold.
In her vision of heaven, the doors open wide,
but the dream quickly fades and she wakes up aware
that the heat was just nonsense – a trick of the mind –
and around her she sees what has always been there
as the bright sunset fades leaving nothing behind.
Grunya sits at the window, her head deeply bowed,
her heart full of sadness, completely forlorn.
But, as night covers all in a dark, gloomy shroud,
the graveyard before her begins to take form.
The stones and the crosses, now vividly glow,
as a silvery light over everything brushes.
Now, dragonflies chatter in grasses below,
and a nightingale sings in the sweet lilac bushes.

In the poem *If You in Life...* there is another apparent rhyming challenge, possibly easier to meet in Russian than in English. In this thirty-two-line poem, only two basic rhyme endings were employed, so that all of the odd lines display one rhyme, and all of the even lines another. This poem seems to indicate a sacred regard for the obligation to use and develop her mathematical gift.

If You in Life...

If you in your life, for one moment, at least,
Have had in your heart a sense of the Truth,
If this ray of the Truth, through the darkest of days,
Has lighted your path, its glow shining through,
Whatever else fate's firm decision has placed
In the path of the future determined for you,
Keep as a treasure held close to your breast
The one sacred Truth, which that moment you knew.
For, though dark clouds may gather in ominous mass,
Obscuring the sky with the dark that ensues,
With a clear sense of purpose and faith's quietness,
Match your strength to the storm and its thunder subdue.
Though false spirits and visions would have you digress
From the pathway that Truth has made open to you,
A salvation against every wicked device
Will be found in your heart when you search it anew.
If you nurture the spark and the Truth's sacredness,
You're almighty and even omnipotent, too,
But beware of the grief, if you let them oppress,
And, before you're aware, steal this treasure from you.
'T would be better with life to have never been blessed.
'T would be better to never have glimpsed at the Truth,
Than to know and then let yourself be dispossessed
And, for one mess of pottage your birthright eschew.
Fearsome gods can be strict, even jealous, on this.
Their judgement, once heard, must not be misconstrued:
The one to whom much has been generously given
Is expected those talents to strengthen and use.
The Scripture is clear on the point of forgiveness:
Man may ask and receive it for all he may do,
Except for the sin against God's Holy Spirit.
That sin God shall never forgive or undo.

In *A Convergence of Lives*, Koblitz says that Kovalevskaya was not at all religious.[52] It is difficult for me to come to that conclusion. The two previous poems seem to have been written by someone who spent at least some time lost in religious thought. I can only speculate that *If You in Life...* may have been written during the time Sofie was struggling with the decision to begin to live separately from Vladimir once again and to focus her thoughts and efforts entirely on mathematics.

A letter from Weierstrass, dated 14 June 1882, indicates she confided her conflicted feelings in him.[53] Weierstrass, who was catholic, had at first felt strongly that Sofie should stay with her husband, but when he understood the nature of the troubles they were having and realized that Vladimir, also, no longer supported her in her mathematical efforts, he decided it might be best for Sofie to go forward with her plans to live apart from him and to continue her work in mathematics, which she decided to do.

Sofie brought Fufa with her and moved to Paris. Vladimir came to Paris for a short time, but the visit did not go well. When Vladimir left to return to his university, they began to live separate lives. Fufa was sent to live with Alexander, Vladimir's brother, where she had cousins her age. Weierstrass had advised Sofia that she might benefit from getting to know some of the French mathematicians in Paris. She met Charles Hermite, who specialized in elliptic functions, who became a friend and advisor. Hermite had trained Henri Poincaré, who also became an important influence in Kovalevskaya's mathematical life.[54]

The next poem is about a very vivid dream and includes no rhyme or regular meter at all, except in a German quatrain within the poem. Kovalevskaya clearly considers the piece to be a poem, however, since she calls it *A Poem in Prose*. The poem is very likely an account of an actual dream. Kovalevskaya is known to have recounted her dreams to her friends fairly often. In her biography of Sofia Kovalevskaya, Anna Charlotte Leffler writes that Sofia often had trouble sleeping and that she would sometimes awaken her and ask her to stay awake with her when she had been troubled by a dream. Sofia is known to have said on several occasions that she thought a dream might be prophetic and Anna Charlotte states that her predictions did sometimes prove to be true.[55]

A Poem in Prose

I had a dream. I was on the shore of the sea – among unfamiliar people. I spoke with everyone – laughing and talking nonsense – so that everybody thought, "What a lively, happy, carefree soul!" Then, everyone went their separate ways and I stayed behind on the empty shore.

The sun had set and twilight was falling. The air was so heavy that it seemed as though at any moment I wouldn't have the strength to breathe.

Some white clouds, like snowflakes, were obscuring the sky.

> *Der Himmel war so trübe,*
> *So schwüle war die Nacht,*
> *So ganz wie unsere Liebe,*
> *Zu Thränen nur gemacht.*[56]

The roar of the sea in the distance was growing closer and closer. I kept going forward. At first, I came across pedestrians who had lagged behind, but, little by little, the shore was becoming emptier and emptier. A fisherman who was walking by told me, "Hurry back, ma'am. The rising tide is getting closer." "I'll make it," I answered, laughing, merrily.

I kept going forward. The roar of the water kept getting louder and louder. The sea extended, unfolded before me, like a steel-gray mass, through which, only here and there, the white crests of the waves would appear.

It was already so dark that it was impossible to tell where the shore ended and the sea began. My feet kept sinking into the wet sand, but I kept going forward.

The wind was blowing in my face. I kept remembering melodies my mother had played for me when I was a child. I was remembering favorite poems. Mathematical theorems were appearing in my head with striking clarity. I was becoming more and more cheerful. I completely forgot where I was and why I had come there.

Suddenly, a huge wave broke right at my feet, soaking me from head to toe. I was at that moment seized by a grotesque, insurmountable fear. In that instant, I comprehended all of the terror of a violent death. I ardently, agonizingly, wanted to live – even if in an unhappy state, in degradation, or in a state of being despised by everyone – only to live!

I started running and began crying – calling for help – but it was too late.

The heavy wave caught up to me and knocked me off my feet. I kept on fighting, struggling against the inevitable – all the while still madly hoping, believing in the possibility of rescue – until a huge wave rolled over my head, quietly whispering in my ear:

Enough!

There will be no more questions on life as before.
There's no need for songs or for tears anymore.

Some of the images of herself in Kovalevskaya's various writings are light and happy, but there is no doubt that she is expressing deep and honest feelings when she declared in the poem *April 13th* that she found no pleasure in the coming of spring.[57] This poem was most likely written after she had experienced *scenes of graves and separation*, possible references to the deaths of her father and sister and her husband's tragic death by suicide. In *Did It Happen...* she tells us clearly about great pain in love, as she writes of her heart's reluctant acceptance of a state of *cold, eternal void* in the face of one man's indifference to her once tender feelings for him. She reveals unhappiness in love again in *A Poem in Prose*, when she writes of *love that arises only for tears*. In one poem, she mentions thoughts of suicide – in another, feelings of desperately, ardently wanting to live, even if in an unhappy state.

April 13th is not the actual title of the poem – just the date attached to what seems to be an unfinished story. No year is indicated. The poem was found among her daughter's papers in 1951. Though it seems to be unfinished, what was written is enlightening, since it is apparently at least somewhat biographical. The poem begins with an introspective assessment of how the coming of spring affects her differently than it affects those she sees around her. It takes the reader through a series of scenes clearly related to the poet's memories of her own life experiences.

The feelings she expresses in the poem may reflect her own feelings at various stages of her life, although she claims not to know who the *hero* and *heroine* in her story might be. The poem ends abruptly with a description of a university town in Germany – most likely an image of Heidelberg, the place where Kovalevskaya first studied university-level mathematics. It is possible that Kovalevskaya had planned to tell her whole life story in epic fashion through this poem that ends with her first experience as a university student.[58]

HEIDELBERG CASTLE   PHOTO BY SANDRA D COLEMAN

The image of Kovalevskaya as a daring young girl who, for the sake of obtaining a passport to Germany to study mathematics, entered into a purely nominal, supposedly non-romantic, marriage, is perhaps informed by the scenes from the poem. Is she speaking of herself and Vladimir when she writes of a young couple ambling through a garden, joyfully making conversation about such abstract ideas as equal rights and freedom, while secretly thinking love thoughts they are too bashful to express? Is she the maiden who, having said her wedding vows, rushes away from her girlhood home with her bridegroom as though a dungeon she were leaving? Surely, she is one of the students who would have viewed Heidelberg as a *veritable heaven*.

April 13th

Spring is here, now, gusting warmly.
The last of winter's cold is fleeing.
The ice is gone; the river's opened.
The Neva's once more flowing freely.
Rain and sun, one after other,
Treat the people passing by us.
They say that, like the April weather,
People, too, can be capricious.

Gala parties, balls, and music,
Only make them weary, lately.
Now they turn to something new to
Occupy their minds completely –
Talk of dachas and the country.
Everywhere I hear their voices.
Fathers adding – talking money –
Making plans and making choices.

It would seem that I'm by nature
Very strange and very foolish.
I regret the end of winter,
For it seems spring's only use is
Painful rousing of the nerves –
A mix of impulses and languor.
I don't find at all enticing
This bright, clear, and cloudless weather.
I'm not one who's looking forward
To the fragrant spring ensuing.
Body wilting – I'm lethargic.
Yet, it seems the blood is brewing.

All these surges effervescent
I'm too weak to be ignoring,
While some worm of doubt, a secret,
Ever at my heart is gnawing.
Always through my head such bleak
And such obsessive thoughts are racing –
Thoughts I do not freely seek –
Scenes of graves and separations.

It seems cruel that the dead are
Always by us being carried –
Body after body headed
For the Smolensk cemetery.

51

Face the window and you'll see
Two more coffins carted by.
You could laugh in disbelief
At how many now have died.

No, statistics do not vainly
Calculate with such precision
That it's in the springtime, mainly,
We make suicide decisions.
Should I find myself desiring
To make end of my accounting
That the day would be in springtime
There should be but little doubting.

All this sun and all this comfort
In my soul such pain provokes!
Surely, Lermontov in spring wrote:
Life is but an empty joke.

Memories of some old story
Come to mind with such persistence.
Who the hero, who the heroine,
Or exactly where the place is,
I am not entirely certain.
An ancient house, very spacious,
A neglected, shady garden
Seem to be a manor gracious.

By the doors to the veranda,
A family's taking tea together.
Air comes rushing through the room —
Fragrance fresh as the May weather.
The woeful samovar is hissing.
A clock upon the wall is creaking.
Round the lighted candle bustling
Butterflies of night are weaving.

Shadows on the walls are passing.
So fantastic and so strange.
From the open hall, adjacent,
Echoes a piano's strain.

Here the hero and the heroine,
Quietly from the veranda,
Come down to the shaded garden,
Hand in hand the paths to wander.

Both so young and both so happy!
Careless laughter rings out clearly.
He, just barely done with schooling –
She, a child! Or very nearly!
Both of them life lures and beckons.
Full of life! Unhesitating!
All its passion and emotion,
They're impatiently awaiting.

Thus, they amble through the garden,
Endless conversation making.
How much joy and fascination
In their talk and their debating!
Thoughts and questions occupy them
Of society and people,
And they find the words exciting,
Talk of equal rights and freedom.

Of themselves, of happiness,
And of love, no word they utter.
Secrets of their youthful hearts,
Bashful lips will not uncover.
Words aren't needed to be knowing
How a happy heart is glowing
Full of life's exhilaration,
When the blood is quickly flowing.

Each, it seems, is hearing plainly
Thoughts the other's heart's proclaiming,
So, without intention, now,
The discourse of the lips is waning.
Thus, they wander without speaking
In the darkness of the garden,
As the coolness of the evening
Fans them, gently, as they're walking.
Through the poplar's fragrant foliage,
Whispers of the breeze are trembling,
While rings forth with silvery trilling
Songs of nightingales, unending.

Now, a very different picture
Opens up before my eyes.
In a church the candles glitter,
for a bridegroom and his bride.
At the altar, joined together,

Having spoken solemn words,
Hand in hand the couple follow –
Led three times around the church.

By the back steps waits a carriage,
Where the newlyweds arriving,
Seat themselves and dash away, as
The clouds of dust are rising.
All the old familiar places
Fade behind them in succession,
As the mountain range obscures
Shady pathways lined with lindens.

An old garden, pond, and groves
Are now fading in the distance.
They see just an ancient turret
Where the waving flags still glisten.

An ancient house upon the mountain,
Drenched in light, would seem to glow
In anger, looking down upon them –
Glowering as it sees them go.

Windows seem like fiery eyes
On a sullen, surly face.
There are no regrets or sighs
As she leaves behind this place.

For, the image of this house
Is not dear to her nor tender –
Only bitter, sad remorse
Does the sight of it engender.

She recalls the pain and troubles –
Years of ardently aspiring –
All her secret, voiceless struggles
And her long-suppressed desiring.

She sees images of slavery,
Curling round her like a string.
She's escaping from the dungeon –
Free at last from everything.

Now, the door is opened to her
To trade acts for aspiration,
And she gazes toward the future
With courageous expectation.

She is not the least bit frightened
Of the road, so unfamiliar –
For her path is now enlightened
By the hopes and dreams that fill her.

Now, unfolds a different image
In the changing scene I'm viewing
Of a tidy German village
With a castle, now in ruins.
All the mountains outlined softly
And the chestnuts lush and verdant
Are enveloped in a misty,
Hazy blue, transparent curtain.

Green, as boundless as the ocean,
And the sky of vivid azure
I see whimsically patterned,
With great arches and with towers.

Nestled down within the valley,
An old, learned town, half-hidden,
Is encircled by its gardens,
As if tied up in a ribbon.

In the daytime, there is no one
On the streets. They are all empty.
But, come evening, of the caps
Of red and white one catches glimpses.
As they pass enormous hounds,
Their loud barking almost deafens,
But to students this small town
Is a veritable heaven.

Not in vain is it referred to
As a breeding ground of science.

I had the pleasure of visiting Heidelberg one summer as my husband and I drove from Nice to Geneva for a meteorological conference. We had researched driving what was at that time called the *Castle Road* in Germany. We arranged to stay overnight in a castle in Bad Friedrichshall on our way to Heidelberg. The images I find now on the internet are not at all a match for the experience we had of staying in a room with antique furniture with locking drawers and cabinets, lace curtains in the bath and the sound of a rooster's crow and church bells as our morning wake-up call. It makes me sad to see that *improvements* have ruined what I saw as so lovely as it was. I wonder if we will be similarly disappointed if we ever return to Heidelberg.

When we were in Heidelberg, I was thrilled to go climbing among the ruins of the castle spoken of in this poem. We enjoyed wonderful jaeger schnitzel at the restaurant where students have been carving their names on tables and walls since the days when Kovalevskaya was there. We enjoyed searching and searching for her name, but did not find it. We did find a photograph on the wall, however, of students wearing the sort of student caps she refers to in the poem.

On another travel adventure, I was able to pinpoint the location of Sofia's apartment in Berlin[59] and the linden-shaded pathway by the river that would, most likely, have been her walking route to Weierstrass's home at 40 Potsdamer Straße during the time he was giving her private lessons.

Joan Spicci, author of *Beyond the Limit*, was kind enough to give me every address she had gathered for places where Kovalevskaya received letters or from which she sent letters. It has been exciting to me to make searching for these addresses part of a number of trips to Paris, Berlin, London, Beaulieu-sur-Mer and more.

ANHALTER BAHNHOF   PHOTO BY SANDRA D COLEMAN

My favorite stop in Berlin was the site of the ruins of the old train station, *Anhalter Bahnhof*, which was bombed during WWII. It is still recognizable as the train station Kovalevskaya would have passed through many times. There was no one there when I visited but me and a man who was just sitting, taking in the site. I think the site is now very different from the way it appeared to me on that quiet morning with only two people present. I have learned that it is now headed for becoming a busy tourist attraction with plans for a museum underway. [60]

Tracing Berlin addresses, I found that Weierstrass's home at 40 Potsdamer Straße no longer exists. A church, *St.Matthäus-Kirche*, and the Berlin State Library, *Staatsbibliothek zu Berlin,* are near where his home would have been. The kind staff at Louisa's Place, the hotel where we stayed on Kurfürstendamm Straße, became interested in my project and eagerly provided former and current maps to show how the locations of even major paved streets were changed significantly after the war.

Kovalevskaya was in Paris many times – sometimes for visits, sometimes living with Aniuta and Victor Jaclard, and sometimes living alone. During some period, she lived in the Latin Quarter, very near where we stay in St-Germain-des-Prés. Our favorite hotel is directly across from Café de Flore, one of the oldest coffee houses in Paris, which opened in 1880. Kovalevskaya would likely have been among the first of a long line of famous visitors as it gained its reputation as a café for poets, artists and philosophers.

ÉGLISE SAINT-GERMAIN-DES-PRES
PHOTO BY SANDRA D COLEMAN

The view from our garret hotel room includes the café and the oldest church in Paris, where we visited the tomb of French mathematician and philosopher, René Descartes.

A POEM BY KARL WEIERSTRASS

Reflecting, now, on the group of nine poems, we see that all of Sofia Kovalevskaya's poems are deeply personal and mostly self-focused. They are based on her experiences, her relationships, her feelings, her beliefs and her aspirations. She has on occasion been criticized for her egocentricity. Indeed, in this respect, as in many other areas, she may well have been her own worst critic.

She confesses in one of the many letters she wrote to Karl Weierstrass, during their twenty-three-year correspondence, that her motive in writing a certain letter to him was entirely egoistic. To this, her gentle mentor responded, "My beloved friend! You began your last letter by saying that it was an entirely egoistic motive that drove you to answer so quickly. To that I could reply, that an egoism that has an origin like yours is in my eyes such an amiable fault, that I honestly wish you might never lay it aside..." [61]

These words were written early in their friendship in April of 1873. It may have been some years before Weierstrass began to regard his student and friend as a mathematician with the soul of a poet, but that she did, indeed, bring to his life an element of poetic beauty, through the words she shared in their long correspondence, is perhaps expressed most clearly in a poetic tribute to women that he himself wrote.

Weierstrass shared this poem in the form of a toast before a large group of prominent mathematicians attending a celebratory dinner during his 70th birthday jubilee. He sent a copy of the poem to Sofia in Sweden with a letter about the party. He wrote very frankly about his thoughts on the mathematicians who did and did not attend. He was very pleased to write that he composed and presented the poetic toast himself, although the first few lines are a quote from a play.[62] Kovalevskaya was not present at the birthday jubilee, but she helped in its organization and with the preparations of some of the gifts to Weierstrass. [63]

Just recently, I obtained a copy of the book, *A Photo Album for Weierstrass,* produced and introduced by Reinhard Bölling, who also published the collected letters of Karl Weierstrass to Sofia

Kovalevskaya. The book is a facsimile of a copper-engraved photo album given to Weierstrass at the 70th birthday jubilee. It contains portraits of students, friends and colleagues, and includes portraits of famous mathematicians. Sofia Kovalevskaya is the only woman included in the album.

In his introduction, Bölling tells about four birthday jubilee gifts to which mathematicians from all over Europe contributed: a marble bust of Weierstrass, a gold coin, a set of two photo albums and an issue of *Acta Mathematica*, featuring a picture of the celebrated mathematician. The photos that were collected included 294 of the very popular carte de visite photos and 40 larger cabinet card photos, displayed in two different custom engraved albums.

Among the guests at the birthday celebration were George Cantor, Leopold Kronecker, Lazarus Fuchs, Hermann Schwarz, Paul Du Bois-Reymond, Heinrich Bruns, William Killing, and Ferdinand Lindemann – some of whom Weierstrass talked about in the letter he wrote after the celebration – all of whom very probably understood that Sofia Vasilievna Kovalevskaya was the inspiration for the poetic toast.

KOVALEVSKAYA AND WEIERSTRASS

Schönheit ist das Weltgeheimniß

"Schönheit ist das Weltgeheimniß,
das uns lockt in Bild und Wort,

Wollt ihr sie dem Leben rauben,
zieht mit ihr die Liebe fort.

Was noch lebet, zuckt vor Abscheu,
alles sinkt in Nacht und Graus,

Und des Himmels Lampen löschen
mit dem letzten Dichter aus."

Also der Poet. Der Forscher,
dem ein güt´ger Gott verlieh

Zu verstehn des Geistes Walten
und der Sphären Harmonie,

Sagt uns: Wahrheit ist die Sonne,
deren Licht das All erhellt,

Und des Wissens Gut das Höchste,
was an Schätzen beut die Welt.

Alles Schönste aber, das des Menschen
sehnend Herz beglückt,

Alles Höchste, das des Menschen Geist dem
Erdenstaub entrückt,

Im Gemüthe edler Frauen ist´s vereint zu
schönem Bund,

Daß uns allen kund es werde
durch der Liebe Zaubermund.[64]

Poem by Karl Weierstrass
Presented October 31, 1885

Beauty is Earth's Sacred Secret

"Beauty is Earth's sacred secret,
luring us by scene and word.

If from life a soul would take this,
with it loss of love occurs.

What is left is seen as garish.
All collapses into night,

And the lamps of heaven perish,
with the poet's fading light."

Thus, the poet. The researcher,
whom our God benevolently

Grants to understand the truth
and the spheres of harmony,

Says: Truth most clearly is the sun,
the light in which all things abound.

Its knowledge is the greatest treasure
which can on this earth be found.

But most beautiful, however, is
that which to our hearts is kind,

Loftiest is that which brushes
all the earth dust from our minds.

Women's gentle, noble spirits
are with beauty in accord.

This becomes apparent to us
through their tender, loving words.

Translated by
Sandra DeLozier Coleman

NOTES ON THE PLAYS

Anna Charlotte Leffler wrote a biography of Sofia Kovalevskaya after her death that was published in London in 1895. It is an unusual biography. Anna Charlotte had actually promised Sofia, repeatedly, that she would write a biography of her if Sofia were to die before her. The biography she wrote claims to portray Sofia Kovalevskaya as she saw herself, which Anna Charlotte points out may not agree with the view of Kovalevskaya that others who knew her may have had. That is not really so surprising in light of the characteristic of Kovalevskaya's personality that the poem *Chameleon* reveals – the characteristic of appearing to be a different person to nearly everyone she met.

ANNA CHARLOTTE AND SOFIA

Many biographers, in assessing the value of the biography Anna Charlotte put together, make a case for dismissing this, admittedly poetic, likely not entirely factual, collection of thoughts related to Kovalevskaya. Anna Charlotte defends her special biography as valuable in a different way than biographies that sketch a person's life history from a more objective point of view. She insists she is telling the story as Sofia wanted it told. She recalls sharing her plan to write a biography with Henrik Ibsen. He immediately asked whether it would be a biography in the ordinary sense or something more like a poem. Anna Charlotte replied that it would be Sofia's "own poem as revealed by herself to me." He totally agreed that this was the right approach for her to take.

Regardless of the verity of every story told in Anna Charlotte's account, there is value in the view of a person who spent more time with Sofia Kovalevskaya than anyone else during the last four years of her life. For purposes of this book, her account of how the plays came to be is probably the best source, since she was there. She was doing the actual writing of the play Sofia was imagining – the play she was wanting with all her heart to finish quickly.

There is much to be learned about the *Struggle for Happiness* plays from Anna Charlotte's memoir of her friend and co-author. We can begin with the origin of the idea to write two parallel dramas.

Anna Charlotte gives accounts of several trips Sofia made to Russia to be at the bedside of her sister, Aniuta, who was suffering from complicated health problems that it was evident would not just go away or be cured. Each time Sofia returned to her sister, she was prepared to be with Aniuta as she drew her last breath.

According to Anna Charlotte, it was during one of those sad and distressing visits that the idea for the plays was born. As Sofia began to reflect on the difference of *how it was* and *how it might have been*, she thought of how different their lives might have been had she and Aniuta not committed a few fatal errors. These thoughts led to the idea of writing two parallel romances. In part one, the fatal errors play out with tragic results. In part two, other choices are made and the ending is happier for all.

Anna Charlotte recorded in her memoir of Sofia that each day after finishing work on the plays they would walk in the woods, where Sofia would shout in exultation, take Anna into her arms, dance around her, and exclaim that life was beautiful! They cherished the most exaggerated hopes for their drama. Anna Charlotte stopped work on a novel she had begun in order to concentrate on the plays. She wrote to a friend that she could imagine their collaborative work having "world-wide success, at least in this world, and perhaps in another."

Although Anna Charlotte says that she normally took months, even years, to begin writing a play or novel for which she had an idea, they worked very quickly to turn out the first rough drafts of these plays. Perhaps they worked too quickly. Their first reading, heard by a group of friends close enough to be honest about their impressions, led to a tedious process of revision. The revisions took much longer than the original writing of the plays and continued through the spring and summer.

Before the play was finished, Sofia was called to St. Petersburg to be with her sister, Aniuta, once more, because her husband, Victor Jaclard, was expecting to have to leave for Paris, on short notice, and Aniuta was too ill to travel or to be alone. The work on the play resumed when Sofie returned to her apartment in Stockholm in the fall, but Gösta was very concerned that Sofie was spending too much time on the plays and neglecting her mathematics.[65] The deadline for the Prix Bordin submission was fast approaching.

He especially chided her for spending time on embroidery. It seems embroidery was helping her stay quietly involved, as she waited for Anna Charlotte to complete the writing for each part of the play, as Sofie's ideas were being unfolded to her. Sofie's command of Swedish was not good enough for her to do the actual writing herself, but, even though she had not yet fully realized that she had talent for writing, she wanted to be part of the writing process at every step. So, she kept her fingers busy holding a needle while Anna Charlotte held the pen and wrote. Anna Charlotte wrote that "while her needle went in and out, her imagination was at work, and one scene after the other was pictured in her mind."

When the plays appeared in print in late December of 1887, they were printed under a pseudonym. Initial reaction was that the plays were more suitable for reading than for performance. The Russian version, which was better-received, was not produced until after Kovalevskaya's death, so she never saw a performance. Still, the plays marked a new phase in her life.

In *Little Sparrow*, Don H. Kennedy gives a very thorough account of the stages of Sofia Kovalevskaya's development as a writer.[66] He tells of early signs of talent for writing being intercepted by life events and by efforts to make significant contributions to mathematics and to the furthering of educational opportunities for all. Her early interest and identified talent for writing were largely neglected until 1887, except for such short pieces as theater reviews, scientific articles and other writings to be found in *Vospominaniya Povesti*.

Written just four years before her untimely death in 1891, the *Struggle for Happiness* plays were the first of her writings to afford Kovalevskaya the label of *writer*. Later, *A Russian Childhood* won much greater acclaim. *Nihilist Girl* was published after her death, but was a success and is still in publication.

THE STRUGGLE FOR HAPPINESS

Two Parallel Dramas

Written by
S. V. Kovalevskaya and A. C. Leffler

Translated by
Sandra DeLozier Coleman

PREFACE TO THE DRAMA

This preface was written hurriedly in rough draft
by Sofia just before publication and given to
Anna Charlotte to edit and write in Swedish.
The following translation is based on the Russian
version that appears in *Vospominaniya Povesti*.

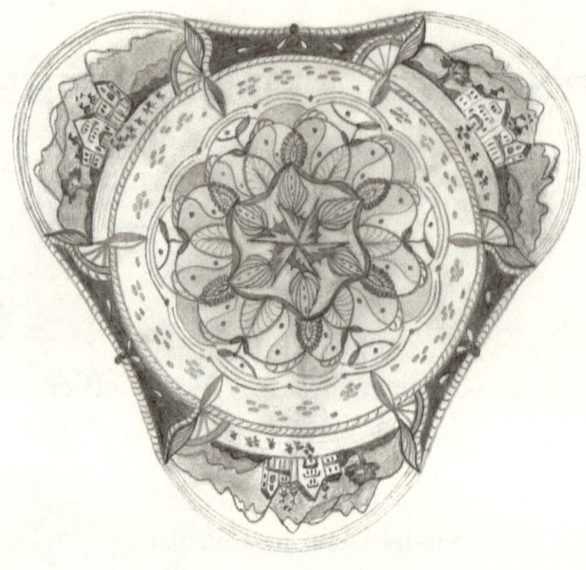

HOW IT WAS AND HOW IT MIGHT HAVE BEEN
Drawing by Sandra DeLozier Coleman

In each of the following plays, the lives of a set of six characters are intertwined. The characters in the two sets are almost, but not entirely, the same. Slight dissimilarities are sufficient to lead them to make significantly different choices at critical moments, leading to very different endings. The image above, like the plot, unfolds from the fires of an Ivan Kupala's Eve celebration. From there, a 60° rotational symmetry pattern represents both sets of six characters, who are understood to have traveled parallel paths until a crisis leads to radically different responses. The resultant opposite endings are reflected in two contrasting, overlapping 120° symmetries. The outer convex vs. concave curves reflect a change in how the characters have been paired.

PREFACE TO THE DRAMA

What person has not thought to question how different his life might have been, if he, in one situation or another, had not made a choice he made in reality, but had made some other choice instead.

When we consider the everyday phenomena of our lives, we always represent ourselves as prisoners of external circumstances. The everyday course of our everyday existence keeps us connected by a thousand invisible threads. We occupy a certain place in life, aware of certain duties, which we execute, exactly, automatically, without straining our forces to the last extreme, and, if we were to wake up in the morning feeling suddenly a little better or worse than before, a little stronger or weaker, a little more or less capable, it would make very little difference.

I could not force the course of my life to change its direction without becoming completely different than I am in reality, without being gifted with absolutely different qualities, which, even in a dream, I cannot ascribe to myself, being always conscious of my individuality. But it is quite different to me, all this, as soon as I bring to mind certain instances in my life. Then, the inherent belief in the existence of free will speaks to me with irrepressible force. It, then, seems to me that it cost me in this or that moment not to stretch my forces a little more, not to ponder more carefully the state of affairs, not to act more energetically, or not to be in some other state of mind, which might have sent my destiny in a completely different direction.

This is similar to the way we regard belief in miracles. There is hardly a man who, without being crazy, would entreat the creator to change the laws of nature for his sake, to force, for example, the dead to return to life. But I will allow myself to ask all believers, who among them has not prayed to the Lord at times for him to make a small change in his orders, to make a patient recover, for example. A small miracle seems to us incomparably easier to perform than a great one, and you really need some mental effort to recognize both of these requests to be perfectly homogeneous. The same goes for our thoughts about ourselves. It is almost impossible for me to imagine how I could wake up unexpectedly in the morning with a voice like Jenny Lynde's or with a body more flexible and stronger. But it would not be at all hard for me to imagine that the tint of my complexion could be different.

It was the way just such a small difference affected a critical moment that the authors wanted to portray in these two parallel dramas. They imagined that Karl in the first drama and Karl in the second would be the same person, except for a few small differences of the sort one might easily ascribe to oneself, if not constrained by consciousness of one's own individuality. In everyday life, we would hardly notice such differences, and, in most cases, they would have no effect on Karl's actions. We should imagine that, if all had gone well, if the hero's father had lived another two years, then, Karl, as drawn in the first drama, and Karl, as he is portrayed in the second drama, would have experienced roughly the same fate. Then all the petty perturbations in their lives that might have been caused by these small, invented by us, differences in their characters, would have disappeared under the pressure of external circumstances.

But then comes a critical moment in their lives, when absolutely identical duties push them in quite opposite directions, so that the small differences in their characters, which we have supposed, are quite enough to make one of them choose one path and the other another. Once the choice is made, the two begin to live very different lives, moving along pathways that never meet again.

Let's take an example from the field of mechanics.

Imagine an ordinary pocket watch, or, if you will, a small heavy bullet, hanging on a very light, but difficult to bend, thread, attached to a nail. If we give the bullet a small push, it moves to the right or to the left, depending on the direction of impact, describes a certain arc, reaches a certain height, and falls back. It does not stop at the place where it was given the initial push, but, instead, moves beyond, in the opposite direction. It then rises to approximately the same height to which it had risen initially, and continues for a certain period of time to swing to and fro.

If my first blow had been slightly stronger, the bullet would have risen slightly higher, but from that point would have continued to move as described above. If, however, my first blow had been so strong that the bullet had reached its maximum height, it would not have fallen back, but would have continued to move forward, making a full circle. As a result, the nature of the movement would have changed. Thus, two strikes, quite similar to each other, one of which does not cause the bullet to reach the critical height while the other does, can lead to quite different results.

In mechanics, we are accustomed to studying these kinds of boundaries of movements, or critical moments, and, sometimes, in order to gain a clear understanding of a known phenomenon, it is necessary to examine it in connection with these moments.[67]

The authors of this drama thought to investigate the impact of this kind of critical moment on two people, very similar to each other, but not quite identical. To understand what the authors wanted to say, it is necessary to remember that the first Karl and the Karl of the second play are not the same person: one of them is more idealistic, better able to distinguish between the important things in life and the insignificant. But these differences are so subtle that in ordinary life we would hardly have distinguished one Karl from another.

If things had gone well, if the father had lived a few more years, so that the son had been able to consolidate his position after his death, the fates of the two Karls would have progressed in the same way. In all likelihood, both of them would have become unassuming scholars, would have become professors at the university, or at a higher technical school, would have married at about the same age and would have made the same choices. But, suddenly, at a critical moment in their lives, a small, barely noticeable, difference between them is quite enough to make one boldly cross the critical point and the other fall under its burden.

The Struggle for Happiness

Two Parallel Dramas

Written by
S. V. Kovalevskaya and A. C. Leffler

Translated by
Sandra DeLozier Coleman

Part I

How It Was

A Drama in Four Acts

CAST OF CHARACTERS

Baron Jacob Yullenyelem	Stenson
Alisa, his daughter	Martha, his daughter
Aunt Amelia, the baron's sister	Mdm Fredgolm
Yalmar	Mdm Selen
Mdm Torell	Gentlemen
Karl ⎫	Lady
Ernest ⎬ her children	Guests, workers, students
Paula ⎭	and others

The action takes place at Gerrgamra, the estate of the Baron Yullenyelem,
and at the Torell home at Lido.

PROLOGUE

The terrace of Gerrgamra castle is visible from the side. There are bunches of flowers, garden chairs, sofas, and a table. This part of the castle has a balcony with a staircase to the right. To the left there is a dark, shady park situated a little below the terrace and set apart from it by a few steps and an iron lattice. Through the trees, factory buildings can be seen. In the background on the left a waterfall is cascading freely at the top, but confined at the bottom by a dam for use by the factory. To the right there is an open view of the mountains, where one can see patches of light from the fires celebrating Ivan Kupala's Eve. The workers, dressed in holiday clothes, begin to appear in the courtyard that runs around the side of the castle extending the terrace. Alisa, dressed in white, is standing to the left of the terrace surrounded by a crowd of schoolchildren, to whom she is giving gifts. Teachers are with the children as they make them approach one at a time, in order. Since Alisa can't see the children in the back of the line, she stands on a bench and proceeds to pass out packages, paying no attention to the arriving neighbors and guests. They are being greeted by Aunt Amelia.

MDM SELEN (*emerges from the castle, turns to Aunt Amelia and points towards Alisa*). What a lovely picture! Everyone's so happy to be welcoming our precious Alisa back home again.

AUNT AMELIA (*seems to be thinking before speaking in a meek, mournful voice*). Of course – it's wonderful! Our precious girl has always brought such life to this house! She has so many plans – plans to organize schools, public lectures, and other projects to help the workers.

MDM SELEN. I heard she performed brilliantly on the university exams.

AUNT AMELIA. Yes, she has a good head on her shoulders. She's like her father. And since the good Lord hasn't seen fit to give my dear brother a son, it's a good thing that at least his daughter…

MDM SELEN. So, she will be continuing with classes at Uppsala?

AUNT AMELIA (*smiling and casting a glance at Yalmar who is coming towards her*). Oh, I'm hopeful that something very different could be happening in the near future.

AUNT AMELIA (*continuing to focus on Yalmar*). It may be that, because they grew up here together, almost like brother and sister, perhaps they haven't really thought about it – but it would please my brother so much. He's always loved Yalmar like a son.

(Yalmar walks up and bows to Mdm Selen.)

MDM SELEN. Congratulations, Yalmar. You must be thrilled to be welcoming home your precious cousin?

YALMAR *(with indifference)*. Oh, yes… of course. But, listen, Aunt Amelia, is it really true that Alisa has invited Martha Stenson here tonight?

AUNT AMELIA. Yes, Yalmar, dear. You know they've been at school together and Martha has become so attached to Alisa.

YALMAR. Does this mean that Stenson will be coming to our house now on a regular basis?

AUNT AMELIA. Martha actually wrote to Alisa saying that she would only be able to come if we also invited her father. What can you do with people who are so out of line? Alisa was very annoyed, but…

YALMAR *(to Baron Yullenyelem, who has walked up to them)*. What do you say to this, Uncle?

BARON *(to Mdm Selen, who is making a questioning gesture)*. He's our neighbor, as well as our competitor. Stenson's the director of the company that owns Lido, you know – that big factory by the city. He's always tried to put Gerrgamra down in every way possible and would put an end to our factory with pleasure, if he could.

MDM SELEN. Yes. Well, might that not actually be a good thing? It seems to me the estate would be better off without all this soot and smoke.

BARON *(smiling and darting a glance at Alisa)*. The future owner doesn't share your opinion. She's very much interested in the factory and its workers.

MDM SELEN. Ah! Yes! Alisa! But won't all her plans be subject to her future husband's plans.

OTHER LADY. Actually, it's my understanding that there are conditions regarding this for transferring the estate to a woman.

MDM SELEN. Yes. Everyone knows that, as successor to a noble estate, she would not be allowed to marry any person who was not himself a noble.

OTHER LADY. Yes, but with regard to her plans for the factory, isn't there also a second stipulation? Isn't she under legal obligation to support the factory no matter who she marries?

AUNT AMELIA. Yes. This is also one of the conditions. Our dear grandfather was like a father to the workers and he was confident that his male descendants would continue our family traditions with regard to the factory and its workers. But he was concerned that, if his daughters were to marry into other families, a husband might have other plans for the factory, so he included this second condition for transferring the estate.

MDM SELEN. Ah! I understand! So, even Jacob isn't allowed to close the factory.

AUNT AMELIA. No, and he certainly doesn't want to, either. Jacob is so caring.
            (*The baron goes towards Stenson and Martha
            and exchanges bows with them.*)

AUNT AMELIA. Look! Jacob is welcoming the Stensons. That's very kind of him!

MARTHA (*sees Alisa and looks at her with delight*). Oh! There's Alisa! She looks like some kind of benevolent fairy standing there among the children!
            (*Alisa nods to her, but continues
            making her way through the packages.*)

BARON (*to Stenson*). So, how is manager Karl? Is it serious?

STENSON. Kaput! He can't even move his right arm anymore.

BARON. It's a great loss for Lido, isn't it?

STENSON (*blinking his eyes*). Do you think so, Mister Baron?

BARON. Certainly. He's such a clever man!

STENSON (*pointing to his head*). Megalomania, invention mania – the serious form, according to the doctor.

BARON. So, you mean you don't have confidence in his great invention?

STENSON. It seems to be pure madness for the most part. It would be better if he were more engaged in the factory itself, instead of ruining the business with expensive experiments. There's really nothing to say. It will be a relief to settle all this business on his son.

BARON. Well, I hear young Karl just arrived home from abroad. Apparently, he's a very talented and well-educated young man. He might be interested in helping his father complete the work on the invention, if it seems to him to have any real value.

STENSON. He's more likely to completely stop being interested, when he sees the present state of affairs.

MDM SELEN (*confidentially to Aunt Amelia*). Wasn't Karl Torell the young man who Alisa…

AUNT AMELIA. Yes, but that was just schoolgirl dreaming. Alisa wasn't even sixteen years old when he was last home. His sister, Paula, and she studied together at school and, there, they talked about him constantly and seem to have made him into some sort of fantasy hero.

(Alisa, having finished distributing the gifts,
leaps from her bench and greets Martha.)

MARTHA (*embraces her*). What a beautiful dress!

ALISA. Just wait till you see what I have hidden under here! (*She shows her a student's peak-cap, hidden under the bench.*) Paula and I have agreed to put our peak-caps on today. But I'll wait until I see that she's really… Ah! Here she is! Finally!! (*She extends her hand to Ernest.*) But where's Karl?

PAULA. He'll be coming soon. He stayed back to be with Papa for a few minutes. There's a lot they need to talk about.

ALISA. What's wrong? (*mysteriously*) Why haven't you put it on?

PAULA. Well, you haven't, either!

ALISA. I was waiting for you.

PAULA. And, now, I'm waiting for you!

(Both take out and show each other their peak-caps.)

ALISA. Ah! Great! All right! Now, we'll put them on together.

PAULA. I only hope they won't laugh at us!

ALISA. Oh, no! Not with two of us!

(They put on their student's peak-caps.)

PAULA. Come on! We'll go in together.

(She takes Alisa by the hand and they join the other visitors.)

MARTHA (to *Ernest*). Now, I'm sorry *I* didn't take the examinations at the university!

ERNEST. Really? Do you think this is going to help them get husbands? Anyway, you're the dearest of all the girls in the world.

ALISA *(drawing Paula aside)*. Well! What has he said? How does he look? Is he still the same as he was?

PAULA. He seems to me to have grown up very much. I even feel a bit uneasy. He seems so resolute, so forceful.

ALISA. Really? Then I'll probably feel uneasy, too. Did he *say* anything?

PAULA. About *you?* Oh, of course! He talked about you, constantly!

(Alisa embraces her. Yalmer approaches and asks to be introduced.)

ALISA. My cousin Yalmar, meet Miss Paula Torell.

MARTHA *(running up)*. Maybe you'd like to introduce *me* to your cousin?

ALISA. Yalmar, meet Miss Martha Stenson. Yalmar's already heard a lot about you.

YALMAR. Yes, Alisa's been telling stories all week about her schoolmates. One evening, I even caught her – may I tell, Alisa? I found her sitting all alone in the dark – crying.

ALISA. Please, stop! I'm not really ashamed of the tears. It just felt so strange to be lonely after being used to having someone around to talk to all the time.

YALMAR. I suggested that she might like to talk to me, but she didn't want to.

ALISA. You didn't understand my mood. *(addressing Paula)* He's so intolerable. He laughs at everything. *(to Yalmar)* You like nothing better than to sit around in a soft armchair with a good cigar in your mouth.

YALMAR. Yes, it's delightful – looking at the blue rings of smoke – which need to be seen! It's better than sitting in darkness.

ALISA. Yes – and you also like growing melancholy after a good dinner while drinking champagne.

YALMAR. Yes, champagne, also, creates a mood.

PAULA. Do you constantly wrangle this way?

AUNT AMELIA. It seems to me that Alisa's being a little unfair. Yalmar never feels so well as when he's at a piano. *(to Mdm Selen and some others)* Really! It's amazing! Yalmar can stay for whole days at a piano, composing.

ALISA. Yes, it's true.

OTHER LADY. It's a pity others can't enjoy your music. You should publish your compositions.

YALMAR. Only to hear them as young ladies would clink them out on utterly worthless pianos.

ALISA. I, for example. Whenever I wish to thoroughly irritate Yalmar, I have only to sit down at a piano and play. You should see him, then!

YALMAR *(to Paula)*. But you, Miss Torell, can actually play.

PAULA. And how do you know this, Mister Baron?

YALMAR. I overheard your music once.

PAULA. Oh? When could that have been?

> *(Karl during this conversation approaches and greets Aunt Amelia and some others and then stands around thinking that Alisa will look in his direction and greet him. But Alisa pretends not to notice him until Aunt Amelia starts talking to her.)*

AUNT AMELIA. Alisa, dear, engineer Karl.

ALISA *(turns around quickly with forced surprise and extends her hand)*. I didn't think you'd be here so soon. Paula says your father...

(Karl is also greeted by Yalmar and Martha.)

ALISA. Oh! We were just talking about music. Didn't you recently participate in a student's chorus in Uppsala? I love student choruses. I attended the concert that was held in Uppsala just after the examinations.

(Karl has been standing near her all this time, but has not been taking in the clues from her eyes and has not really been listening to what she's been saying).

KARL. You were – you were at the concert at the cathedral?

(Alisa continues to look at him with wide, revealing eyes, but doesn't answer. Karl, feeling confused, turns towards Yalmar.)

ALISA *(pulls Paula aside)*. You won't believe how amazed I was when I heard his voice just now.

PAULA. Why?

ALISA. Don't you see? It's so hard to explain – but for the past three years, even though I've held his appearance, his glance and his manner clearly in my mind, I never gave any thought to his voice. Just now, when I heard him speak, everything suddenly became so *real* – so stunningly, thrillingly real!

(She embraces Paula and they return to the group.)

YALMAR *(to Paula)*. Would you like me to tell you, Miss Paula, about hearing you play? It was three years ago, just after the new organ in the chapel at Lido had been received. One autumn evening, going to chapel, I noticed that the entrance door was standing wide open and that music was flowing from inside. As the music was not bad, I entered. Suddenly, it seemed as though I had been transported to one of the small villages in the dark Urals. I heard from the pew a bounty of pretty, maidenly musings. I should say, by the way, how unusual it was to be hearing these sounds in a church. It seemed to me, sitting there, that I was overhearing someone's deepest secrets. Never before has anyone told me so much of what young girls say when they dream. That's how the music sounded.

PAULA *(embarrassed)*. And what exactly did you learn from this music?

YALMAR. Well, I'm sure if I were to tell you, you would begin to contradict me. All girls are hypocritical in matters concerning such questions. Are you, by any chance, familiar with this verse from Runeberg?

> All maidens in practice love kisses,
> though they may despise them in words.

PAULA (*turning away from him*). I don't know why you would want to ask me this, Baron.

(*During this conversation Alisa and Karl have been avoiding one another, although, while talking with friends, they have been throwing each other stealthy glances. Suddenly from the courtyard a shot is heard in the distance.*)

AUNT AMELIA. Ladies and gentlemen, please come with us to the other party. We've just heard the signal! The Kupala's Eve festival has begun!

(*All dash into the courtyard. Immediately after that,
fiddles are heard playing as festive dance music begins.*)

ALISA (*gives Paula a sign to approach*). Stay here. I need to talk to you.

(*Soon, Alisa and Paula are left alone on the stage.*)

ALISA (*throws her arms round Paula's neck*). Oh! Paula! I'm so miserable!

PAULA. Why? What's wrong? What's wrong with you?

ALISA. It seems I was silly to believe in dreams!

PAULA. What's happened?

ALISA. Haven't you noticed how terribly disappointed he is in me? And I, too, am terribly, terribly disappointed!

PAULA. Really? Has something about him changed so much?

ALISA. Something about *him*! No! Instead, *he* thinks that *I'm* horribly changed. Can't you see how he avoids me? Oh! He's absolutely, entirely different from the way I imagined him. And now I feel such a terrible emptiness. In fact, I could just die!

PAULA. He's just feeling a little unsure of himself. That's all. I'll talk to him.

(Paula approaches Karl, who has left his secret corner a couple of times during this conversation to look over at them. Alisa begins to converse with some other visitors, staying away from his secret hiding place.)

KARL *(to Paula, confused)*. Well, what's she been saying? I can see that I'm completely different from the person she drew in her dreams? It should have been expected. She was only fifteen years old the last time we saw each other.

PAULA. So, you *do* find it very great – the change in her?

KARL. Oh! Yes, definitely! There used to be something gentle and pensive in her. Now, she's as cold as ice.

PAULA. That's your fault. Why don't you talk to her? She's very much... Your impression on her is just as strong as before. She's just close to tears because you haven't paid any attention to her.

KARL. What are you talking about? I haven't paid any attention to her! She was never so charming as she is now. She's a thousand times better than before.

(He gathers the courage to approach Alisa.
She immediately turns around towards him.)

KARL. Miss Alisa – Paula says... *(Alisa ceases her stern look.)* Paula says you're very much... interested in my father's invention.

ALISA. Yes! Yes, it's true. It interests me very much – very, very much. Would you tell me about it? I want to hear every detail.

(They go together to the left and disappear.)

YALMAR *(emerges from the festivities in the courtyard with Paula, stopping up his ears)*. We have to get out of here! We have to get away from these horrible scraping sounds! That fellow must be drunk or must have lost his mind. If things were as in the old days, I would order someone to throw him in a cellar and lock it up!

PAULA. Be happy it's a fiddle he's playing and not a harmonica!

YALMAR *(stops suddenly)*. A harmonica! Oh! No! Their impudence hasn't gone so far as to torment my hearing with such a tool as that! We really have to get away from here. We can at least try to muffle this scraping sound with good music. Let's go back to the house. I have an excellent grand piano and I'd really like to see whether you can manage to accompany me if I play a violin. None of my neighbors could learn to do it.

PAULA. You must be terribly demanding!

YALMAR. Demanding! You say! Actually, I have been known to fly into a rage when my accompanist plays lousy music. But let's go inside and give it a try. If it's true that you're as talented as it seems you are, I may even teach you a few things.

PAULA. Oh! I can't imagine anything more pleasant than to play for you. But at the same time – I'm awfully afraid.

YALMAR (*entering the house*). Well, then, this first time, I'll try not to be quite so demanding – while you're getting used to me.

MARTHA (*accompanied by Ernest comes back from the dance floor*). Look! Paula is leaving with Yalmar! And Alisa is leaving with Karl!

ERNEST. And you're staying here with me.

MARTHA. I? Oh! Nobody pays any attention to *me*!

ERNEST. Well, as it's no shame to you, I'll just say it! You've grown prettier since last winter! In fact, I don't understand what's happening to me right now.

MARTHA (*flirting*). *What is* happening to you?

ERNEST. What's happening to me! Even this she says so sweetly! No, you truly are too, too appealing! And had I already gone as far as Karl, if I had already graduated from the university…

MARTHA (*with lowered eyes*). Then, what would you do?

ERNEST. I would marry you! Oh! You little sorceress!

(*He embraces her.*)

MARTHA (*offering her lips for a kiss*). As long as nobody sees us!

(*They enjoy a long kiss.*)

MARTHA. Should we announce it before you return to Uppsala?

ERNEST (*releasing her from his embrace*). Announce it?

MARTHA. Can I at least tell Paula and Alisa about it?

ERNEST. Tell Paula and Alisa? About what? That I kissed you?

MARTHA (*patting her lips*). Oh, you're just teasing me! Tell them that we're engaged, of course!

ERNEST. Engaged! There's nothing to tell! Some story! First-rate student has become engaged! (*laughing*) It would be interesting to tell that to Uncle Stenson! No, no! It's better to wait a little, or Papa won't finish his invention.

MARTHA (*now pushing him away*). Well, dearest Ernest, I can wait. Don't be thinking *I'm* in any hurry.

ERNEST. Now, you've taken offence! Don't do that to me! It's just these Ivan Kupala fires and dances – all this has affected me so strongly that I've nearly lost my mind. On an evening like this, it would seem to me that it ought to be permissible to dream a little – permissible to be happy, without having to think about the future. Oh, Martha, don't be angry!

MARTHA. But, then, what do you want? I, truly, don't understand.

(*She begins to understand.*)

MARTHA. Well… All right. I'll let you kiss me again – only not here. Over there, under the trees, because, if you don't want us to be engaged, surely you must understand…

ERNEST (*takes her by the waist and withdraws into the depths of the park on the left*). Such a child. Such a sweet, innocent child!

MARTHA (*as they are leaving, says again in a pleading voice*). Ernest, my love, please – just Alisa and Paula.

(*Karl and Alisa come back and sit down on a bench in the park.*)

ALISA. Tell me everything you have planned!

KARL. It would be better if you'd tell me about yourself, rather than just listening to me all the time.

ALISA. But I really would like to have a better understanding of your plans for the invention. Do you feel confident that your father is right – that it can be a success?

KARL. There's no doubt about it. And I'm so happy that, now, I'll finally have a chance to work with him on it! When I think of how I've dreamed of this moment since I was a child! Ever since my father first explained to me how much force goes to waste during the spring and autumn floods and what a strong push the industry would receive if it were possible to harness the force of the water, the falls have had a very special importance to me. Every time I hear the roar of the rushing water, it's like a voice inside of me echoes the call. Yes! Yes, the day when the battery will finally be put to work will be a day of tremendous celebration for me!

ALISA. But, it's sad that you're returning to find your father so ill and his condition to be so serious!

KARL. Actually, his spirit, his mind, and his capacity to work are still enormous. I see him sitting in his armchair, half broken by paralysis, and, yet, still developing such perceptive, ingenious ideas. I really don't have the power to express how amazed I am.

ALISA. But, if…

KARL. If… You're thinking, what if his illness should take a dangerous turn – if he should die, not having finished the project? Yes, the thought haunts me. All these nights, since I've been back home, I've not been able to sleep. Sometimes, in the middle of the night, I get up and go to his door and listen, worried about another stroke.

ALISA. But, in any case, if one day you were able to finish the work he's begun, you'd find consolation for yourself in that.

KARL. Yes, but if this misfortune were to happen soon, I don't know if I would be in any position to complete the work.

ALISA. Why not?

KARL. I can't expect you to understand, Miss Alisa, but to finish such a project through to the end requires a great deal of money. My father receives a large salary, and this money has sufficed until now, but, as soon as he dies, there will be enormous difficulties. But, yes – certainly – by all means, I…

ALISA. Yes, but when, at last, you *do* finish, you'll be very rich. Isn't that true? Won't you be extremely wealthy then?

KARL (*smiling*). So, that's what you're thinking! Well, yes, then it would even be possible for us…

ALISA. I already have in mind a wonderful plan for how you should use the money.

KARL. And that is?

ALISA. You will, certainly, laugh. But it's such a good plan! You've read about the settlement houses in London?[68] We could arrange something similar here for Lido and Gerrgamra.

KARL. So, this is what you think and dream about. I must admit that I've indulged in more personal dreams of the future.

ALISA (*looking down*). Well, I, too, have had personal dreams.

KARL (*takes her hands*). You should tell me about them.

ALISA. I have dreamed, that you would help me and advise me – and that we would work together.

KARL (*bending closer to her*). I'm so happy to hear these words! But your cousin, Baron Yalmar, wouldn't he be...

ALISA. What?

KARL. Wouldn't he be jealous of you and me? He's probably become used to being your closest friend and adviser.

ALISA. No, of course not. Yalmar's like a brother to me – but you don't always tell a brother what interests you most. He – jealous of you and me?! Definitely not! Yalmar is absolutely never jealous! He's too indifferent for that.

KARL. He's indifferent?! He lives with you, constantly – every hour, every minute, he's near you, and...

ALISA. Yes, but he's not interested in anything but his music. It's like there's a wall separating him from the rest of the world. He's completely unlike me in this. There's something strange about my character. I might seem to be very capable, but, at the same time, I know that nothing will ever come of me, unless I have a person in my life who shares my interests completely.

KARL (*kissing her hands*). That person is right here beside you.

ALISA. Yes, to be sure, I believe you could be this true friend to me! And you would then discover how I could be to you a better friend than you could ever have imagined. I'm sure you have no idea of the state I can go into in my interest in anyone I love – how all my ideas, all my feelings, all my spirit can become focused on one purpose: to live a full and complete life with that person.

KARL. Oh! When you talk like this… but for the time being it's impossible.

ALISA. It's impossible?

KARL. But it will be possible, someday. Only, have the patience to wait for me! Until I'm a success!

ALISA. What are you talking about? We would be happy just to have each other – to be able to be together every day.

KARL. Oh, you don't know – you don't understand. I'm in no state to live near you without wanting more – more every day.

ALISA (*bending towards him*). And do you not think I could give more – more every day?

(Karl draws Alisa to himself, whispering to her. Baron Yullenyelem comes into view in the depths of the stage and sees them. That same minute Yalmar and Paula appear, going down into the park, passing under the trees. Yalmar takes her by the waist. They go further, leaning their heads together, and almost collide with Ernest and Martha, who are walking towards them, also having embraced. There are general exclamations of surprise. Martha and Paula embrace each other. It is twilight.)

ERNEST. There seems to be some kind of magnetism in the air tonight. On a night like this, I think it would be impossible for any person with a drop of blood in his veins not to be a little bit enamored.

ALISA (*runs up to them and embraces Paula*). What a wonderful night! Even the air! How intoxicating! How perfectly beautiful the sky and the hills, there in the distance. Oh! Paula! It's so good – so completely wonderful just to be alive!

PAULA (*whispers*). So, now he likes you?

ALISA (*stepping aside*). Oh! Yes! (*embracing her once more*) I know one thing for certain, that if I searched the whole world I couldn't find another sister as lovely as you.

MARTHA (*also embraces Paula and distracts her from Alisa*). Yes, such a lovely sister – I agree, because sister sounds so much nicer, than sister-in-law.

ALISA (*eyes wide*). Sister-in-law!

(Martha nods her head, significantly.)

PAULA. Oh! Alisa! Surely, you're joking! In all my life, I've never heard the likes of this! What an imagination you have!

BARON *(strolling back and forth near them, approaches Alisa with a displeased countenance)*. I don't doubt that you're all very merry here together, my young friends, and that you will find me terribly prosaic if I dare to remind Yalmar and Alisa that they are completely forgetting about our other visitors. *(Quietly, in a stern voice, to Alisa)* I am surprised at your behavior.

(Alisa immediately leaves and approaches the visitors at the back of the stage. The other young people disperse and mingle with the invited crowd, except Karl, who is approached by the baron.)

BARON. I need to have a talk with you, Mister Engineer.

KARL. To explain…?

BARON. Is it possible that you've dared to abuse my hospitality and trust to compromise, to some degree, this young girl, my daughter.

KARL. I know that, for now, I would be considered to have no right, but I have hope, that in time…

BARON. Perhaps you don't know about a condition affecting the inheritance of Gerrgamra prohibiting the successor from marrying a person who is not a noble? Do you not understand how you would deprive my daughter if you should manage to win her love?

KARL. I would deliver to her, in time, another Gerrgamra to replace what she would lose on my account.

BARON. Well, I can see that you are not lacking in self-confidence. Your dreams are very far reaching. But, I should think you would wait until the time you have actually achieved this brilliant position before ignobly taking advantage of her gullibility. Remember, she is little more than a child. I have no doubt that it is easy to convince her of the feasibility of such dreams, but…

KARL. It never entered my mind to make her any promises.

BARON. I take you at your word. But I must ask you to promise to hold off for the time being – to cease for a year from sharing any private conversations, any secret meetings, any correspondences.

KARL. I have no desire to be accused of trying to gain her favor by deceit. But, if in a year she should decide...

BARON. We'll discuss that in due time. For now – play the game honorably. Give me your hand on it!

KARL. I never thought to play differently than honorably.

BARON. But, as I wish to be equally honest, I must tell you, frankly, that I am hoping that, before a year passes, Alisa will have made another choice.

KARL. In that case, I'll not complain. If she loves me, she'll wait. If not...

BARON. There's no call to attach so much value to the first flash of love of an eighteen-year-old girl. Alisa loves Gerrgamra so much that she'll not be able to leave it behind easily. Believe me, it would be a terrible sacrifice for her.

KARL. Rest assured, I'll never demand any sacrifice of her.

BARON (*extends his hand*). And this is right. (*He then goes to the back of the stage and gives Yalmar a sign to approach.*) Now, Yalmar, tell me what I should think concerning your behavior today?

YALMAR. I don't understand, Uncle.

BARON. First, you disappeared with this young girl – just the two of you together – and then, pardon me, but I was nearby when you came out of the forest.

YALMAR. She has such unusual musical talent – this alone...

BARON. Well, in that case, she's the perfect match for you. You two can drive about together every day and give concerts to earn your daily bread.

YALMAR (*being dared*). Yes, that would be a perfect life for me.

BARON. In the meantime, all this has greatly upset me. You children need to understand that I cannot endure this. My illness makes me very vulnerable and also irritable. (*He takes him by the hand, presses the other hand to his heart and sighs deeply.*) When I think that, as my descendant, I had planned to leave all that I have in your hands – for you know that I love you no less than I love Alisa – and now both of you, on the basis of casual whims, want to turn all plans for the future of Gerrgamra upside down.

YALMAR. Oh! As for me, the situation is not that serious. But with Alisa, it is perhaps worse.

BARON. But why do you pay her so little attention? Isn't it true that you love her?

YALMAR. Certainly, I love her as much as I can love in general, by nature. But she's absolutely not interested in me. *(He tries approaching Alisa.)* If you'd like, we can go and dance.

ALISA *(with anxiety, looks back at Karl)*. No, thank you. I don't feel at all like dancing.

YALMAR. Well, actually, I don't either. Today has been so devilishly complex! *(He leaves.)*

BARON *(to Alisa)*. I expected from you, my friend, more self-control, something better than a woman making advances.

ALISA. What have you said to him, Papa?

BARON. He appears to be very sensible. He had not intended to commit himself at this time. He is likely much surprised at the haste with which you were encouraging him. If you only knew what a young man imagines deep down in his heart – always without fail they have these images – if a young girl throws herself at him too soon. Only restraint and modesty in a woman can be seen as reputable to a man. A woman should never show him all the force of her feelings! A man must not have been assured in advance of the answer a girl will give when, at last, he's prepared to make to her an offer. When I asked your mother to marry me, do you think I knew in advance what she would say to me?

ALISA. These are all empty, senseless, false ideas about propriety!

BARON. Well, then, go to him now and ask him to marry you, if you want, but be assured that you will be refused.

ALISA *(leaping up)*. Oh! Papa! *(She runs from him to Paula, who is approaching with music in her hands, and grabs her hand.)* Tell me! Is it true, Paula? Papa says I've shown too much.

PAULA. Why not show what you feel? *(addressing Yalmar, who is approaching them)* I can see that I am going to need much, much practice. When should I come to play with you again?

91

YALMAR (*troubled*). When…? Truly, I don't know. I can ask Alisa to write and tell you. It's probable that some time will pass, before…

PAULA. Some time…? You had talked about next week.

YALMAR. Yes, I had forgotten that I'm very busy, now.

(*Paula looks at him, absolutely amazed.*)

YALMAR (*takes her hand*). Look – fires die away and become damp and cold. This always happens to the dreams of the Eve of Ivan Kupala. It's impossible to rely on them. It's no use to trust them.

PAULA. Oh! (*He departs from her. She stands, completely stunned. The music falls from her hands. Karl approaches Alisa to say goodbye.*)

ALISA. You're leaving – already – and in such a hurry?

KARL (*darting a glance at Baron Yullenyelem standing nearby*). Yes, your father has brought to my attention that (*lowering his voice*) much time will pass before I could expect to have the right to speak with you the way I spoke tonight.

ALISA (*also speaking quietly*). But you will come to see us again, soon. You'll come often. Won't you?

KARL. I'm afraid it will not be frequently.

(*He shakes hands with her quickly and leaves. Alisa stands amazed as her eyes follow him. Then both girls, in tears, fall into each other's arms.*)

ACT ONE

The hall at the Torell home is cluttered with furniture that has been prepared for a sale. Mdm Torell bustles about, hanging tags on furniture, setting prices. Paula, Ernest and Karl sit at a table having breakfast. All are in deep mourning.

MDM TORELL. Can't you give me a little help, Paula? You're sitting at breakfast too long. I'm afraid the buyers will begin arriving soon. We advertised in the newspapers for 9 o'clock.

(Paula and Ernest rise.)

PAULA. But isn't everything ready for them, Mama?

MDM TORELL. No. No. You sit, sit, Ernest. You and Karl need to eat properly, but, for me and Paula, it's not so important. Here, help me, darling.

PAULA *(runs up to her)*. What, Mama?

MDM TORELL. Look here. The price isn't marked on this chair. Oh, it's impossible to rely on anyone! And now they'll be starting to come! Quickly! Get some paper and a pen.

(Ernest runs for paper.)

PAULA. But, Mama, dear, the price is on the sofa: a sofa and six chairs – 100 krona.

KARL *(continuing to have breakfast with nervous haste, rises and throws a napkin on the table)*. I'm done! You can clear the table, Paula.

ERNEST. I'll help.

(Both clear the table.)

KARL. Mother, I asked you before if it wouldn't be better for you to go away somewhere and have Paula and Ernest stay and conduct the sale? You'll only torment yourself needlessly here.

MDM TORELL. No, Karl, my love, I ask you not to say this. I wouldn't have a moment's peace. I have puttered with this furniture for so many years that I have no wish to leave now while I can still call a single chair my own. Just think how carefully I protected all this as I coordinated all the household affairs during your Papa's lifetime. And, now, all these things will be ruined!

KARL. Can furniture really have such value? These are just trifles!

MDM TORELL. It's easy for you to speak of trifles. You were gone from home for such long periods. You didn't live with all these things, day by day, as I did. As I protected the furniture, I wiped the dust from every little thing myself. *(She begins to cry.)* Maids are always so awkward. I always used only knitted cloths for dusting, cloths knitted from the finest thread, so that there's not a single scratch on any of the furniture. At least you can't reproach me, saying that I was irresponsible in keeping things in order and in protecting our family's property.

KARL. Mama, dear, we know all this perfectly, but why talk about it now? The only thing I can think about, just now, is how to find the necessary means to finish Papa's work.

ERNEST. Oh! Money always finds a way into the picture, somehow.

PAULA. Just wait and see! Someday, some rich man – one of Papa's old friends – will show up – ready to lend you all the money you need to continue his work. I'm sure of it – absolutely sure.

KARL. You're a dreamer – just like Papa. "So, what if I run up a bit of debt," he would say. "As soon as my work has made even a little progress, all the rich people will compete to support me. And how has that turned out? He's died, not having finished.

PAULA. That's all the more reason to trust that the future can't be so bad. One day, we'll all move to Stockholm. You and Ernest will work at supervising the opening. I'll be able to study in a conservatory, and Alisa will be able to attend the university. And she'll live with us! Oh! What a wonderful life we'll begin to live – all of us together!

KARL *(smiling)*. And if this rich man doesn't exist – I say *if*, just because at the moment no one comes to mind.

PAULA *(swings her head significantly)*. But I know he exists – to my mind there may even be two benefactors – one for you and one for me. *(She has put a pawn in her hands behind her back.)* Tell me, which hand do you choose – right or left?

KARL *(smiling)*. Right.

PAULA (*stretches her right hand and opens it*). Well, so you get Baron Yullenyelem.

KARL (*becoming gloomy*). Most humble thanks, but sooner than agree to begin a state of dependence and become indebted to Baron Yullenyelem, I...

MDM TORELL. But Baron Yullenyelem was always a kind friend to your father.

KARL. I know that, but I could never go to him for help. And who did you have in your left hand?

PAULA. Alisa!

KARL. Alisa? What does that mean?

PAULA. Alisa would be glad to lend me the money to continue my studies.

KARL. But Alisa is still a minor. Nothing is hers to give – which means, it would still be Baron Yullenyelem.

PAULA. Yes, but he'll do anything Alisa wants.

KARL. Well, at least, don't call it a loan. Call it relief aid for the poor. Because hopes for repayment...

PAULA. When I become a famous pianist or composer, I'll repay all our debts.

KARL. These are only dreams. Dreaming can be very pleasant. But, from Papa's example, it's clear what results.

MDM TORELL. Again, he reproaches his father!

KARL. Not all my thoughts are reproaches! No one knows as well as I do what a clever person Papa was – what great value there is in the business on which he was working. I also know that I can, and eventually will, finish his invention. But, for now, our main concern is answering the pressing question of how we're going to maintain our existence? And so, I object to the fact that Ernest and Paula are wasting time absorbed in impossible dreams.

ERNEST. Well, what, then, in your opinion, *should* we be doing?

KARL. The same thing I'm doing – advertising in newspapers, going around everywhere searching for work. It's simply a waste of time to sit at home, feeding on hopes of completing the invention on the basis of dreams of platonic love. We have to put the effort to earn money for completing the invention above everything.

MDM TORELL. I hear a call! Oh! Lord! I've forgotten to put a broom in the lobby for brushing feet. It's raining in the courtyard. Paula! Paula!

PAULA. You've forgotten, Mama. I remembered the broom this morning and Ernest himself went to get it.

MDM TORELL. Well, that's good. It would cost a fortune to have this huge floor washed.

KARL. Look after Mama, Paula. See that she doesn't get too tired. I'm leaving.

MDM TORELL. Where are you going, Karl, dear?

KARL. I don't know – to the highway – to wander aimlessly – anything to avoid seeing all these people.

MDM TORELL (*throwing her arms around his neck*). Karl, are you so miserable?

KARL. I'm not into this business and it's so unpleasant to have everyone stare at you.

MDM TORELL. My poor, dear boy. You *are* miserable. And after so many successes and honors.

(*While they talk, Paula leaves to open the door and comes back with Martha.*)

MARTHA (*approaches Mdm Torell and offers her cheek for a kiss*). It's such a pity – such a pity! (*She extends a hand to Ernest and Karl with lowered eyes, then embraces Paula.*) Show me, please, anything here that you'd like to keep. Papa has given me money to buy anything that's dear to you – for you, certainly, not for myself. I'll buy it, and then I'll give it to you as a gift. (*She looks around at the things for sale.*)

(*Karl is trying to leave.*)

MDM TORELL (*aside*). Wait a bit, Karl, dear. It's not necessary to be impolite to the Stensons.

KARL. What! Are you saying I should stand here and accept charity from this little fool?! No, thank you!

(He tries again to leave.)

MARTHA. Are you leaving, Mister Karl? Actually, Papa asked me to tell you – how forgetful I am – Papa asked me to tell you that he needs to speak to you. He'd like you to come to see him.

KARL. When? Now?

MARTHA. Yes – well – you can wait a bit.

KARL. No, thanks. I had better go now.

(Leaves hurriedly, but meets Mdm Fredgolm at the door.)

MDM FREDGOLM. And, here's Karl! Hello, Ada, dear! How nice to see Karl at home at last! Look how he's grown! Look how shapely he's become during his absence. Yes, Karl, at this difficult time it's important that you've come back home! My Fredgolm used to say of your father: Karl is an ingenious person, but he doesn't understand life. That's what he always said.

KARL. Had my father lived a few more years he would have shown that he understood...

MDM FREDGOLM. Yes, but, as Fredgolm says, one must manage one's affairs in such a way as to always be ready to die. That's what Fredgolm says. Also, I admit that, for me, it's always been a great consolation to think that, if the Good Lord needed to take Fredgolm from me at any moment... *(looks around)* Oh! My Lord! How I remember the wonderful hours we spent here in this house. For example, last Easter for dinner. Torell sat there, like a kind angel, so meek and absent minded, in a new velvet cap. I need only close my eyes and I can see him in front of me as if he were still alive. Here, there was a buffet, and, there, a piano. Oh, my word! How awful this is!

(She and Mdm Torell begin to weep.)

PAULA. Aunt, dear. Please don't upset Mama.

MDM FREDGOLM. You're right, my child. What was it I really wanted to say? Ah, yes, here it is: Is there not something here, Ada, that I could buy for you? I would willingly do you that service.

ERNEST. There certainly, is, Aunt Fredgolm! The grand piano, for example.

(Karl leaves.)

MDM FREDGOLM. No, my friend, that's too expensive for me.

(Other buyers appear. Paula and Ernest begin showing them the things for sale.)

MDM FREDGOLM *(pulls Mdm Torell aside)*. When I read the announcement in the newspaper, I told Fredgolm, I know Ada Torell. I understand, how painful it is for any decent person to think that her things... Karl has left?

MDM TORELL. Yes, it's too painful for him to be here for this.

MDM FREDGOLM. The poor boy. Yes, I'll speak to the boy, because he was with my Freddie for a year. Good heavens! I said to Fredgolm, I know how Ada always cherished her things, and, now, I said, when she's not in any condition to be paying debts, there's one remaining.

MDM TORELL *(timidly)*. What debt?

MDM FREDGOLM. It's their business, the men's. Fredgolm will talk to Karl.

MDM TORELL. Poor Karl. He's already found out about so many new debts during the inventory of the inheritance. If only it were possible for us to make ends meet, so that we could finally pay off all the creditors.

MDM FREDGOLM. But Fredgolm, for the time being, is in no hurry for the money. He says that, although business matters are somewhat confused, there are great resources available in the inheritance. The whole debt is secured by the electric machines and various other devices.

MDM TORELL. Yes, but it would be so hard for Karl to sell them.

MDM FREDGOLM. It might actually be to his advantage. At least, he wouldn't be tempted to continue his father's experiments. And meanwhile, dear Ada, if you have anything especially dear to you that you would like to see in the hands of decent people...

MDM TORELL. Ah, yes, perhaps my little chest with the inlaid wood. You see, the pieces of wood fall out so easily and it's so difficult to find the same kind of wood. I admit, it would be very difficult for me, if I should see it somewhere and it should appear that, here and there, there was missing wood.

MDM FREDGOLM (*passing with Mdm Torell into another room*). Think, dear, how it was for me to look at this mismatched set every day – one cup with wider gold band than the others – and just because some negligent servant...

(They disappear on the left.)

PAULA. Martha, tell me! Do you know why Karl needed to see your father? No, don't say anything! I would rather guess. He wants to offer him money on loan to continue working on the invention?

MARTHA. No, this is about a position – about a very good position.

PAULA. This is what I had predicted! Mama, Mama!

(Mdm Torell and Mdm Fredgolm appear in the doorway.)

PAULA. Karl is getting an excellent position.

ERNEST. And then we can all live together and continue to work.

MDM TORELL. Oh! My Lord! My head is spinning! Excuse me! I need to sit down.

MDM FREDGOLM. Yes, I dare say that the Torell family is extremely lucky.

ERNEST. Should we thank you, Miss Martha, for this? Would you thank your father for us?

MARTHA (*embarrassed*). Oh! That I can do! *(to Paula)* Show me, quickly, what you would like to keep most of all. The piano? Yes, clearly. Unfortunately, I don't have enough money. Wait a bit. I will talk with Father. Your grand piano really must not be sold. It would be such a pity. You play so well!

PAULA. Oh! If it were possible to avoid selling it, I'd be so happy!

MARTHA (*cries, being deeply moved at herself*). It will be yours, even if I have to sacrifice all my pocket money.

PAULA. I felt sure this would happen. I was sure that when the piano was just about to be sold... Only, I imagined it a little differently. I imagined that a man – someone absolutely unfamiliar to me – some traveler – would come, would sit down at the piano, would try it and would say that the piano was good. Then I would stand beside it and burst into tears. And then...

ERNEST. And, now, Martha has burst into tears instead of you. *(He takes out a handkerchief and wipes Martha's eyes.)* Allow me! Well – now, I shall never have it washed. How does that story about Peri and the angel go? A tear from an innocent heart... Doesn't it open paradise? [69]

(Karl enters quickly.)

ERNEST. Well?

(Karl makes a sign that he can't speak with Martha there.)

MDM TORELL *(sees Mdm Fredgolm to the door)*. Thank you, dear! You may tell Fredgolm, that I hope – that is, if Karl should now have obtained a good position...

MDM FREDGOLM *(to Karl, extending her hand to him)*. I am so very glad for you.

KARL. About what?

MDM FREDGOLM. That you will be getting a position at Stenson's. Yes, I assure you, I'm pleased, and absolutely not for myself – because Fredgolm, in fact, can wait. You don't need to worry about this debt. He says that it's sufficiently secured by all the devices.

KARL. I don't have any position, yet.

MDM TORELL. Perhaps you haven't seen?

KARL. No, I saw Stenson.

(He looks again at Martha.)

MDM TORELL. Girls, go, please – to Paula's room.

(They leave.)

MDM TORELL. Well, Karl, dear? Don't hesitate to speak in front of Aunt Fredgolm. She wishes us well.

KARL. Stenson did offer me a place, but under a condition to which I cannot agree. He requires a commitment to discontinue my father's experiments.

MDM FREDGOLM. This is excellent news! I find it very clever of Stenson. Fredgolm always said that it was the invention that had ruined Karl.

KARL. Excuse me, but it's not for Uncle Fredgolm to judge.

ERNEST. What position has he offered you?

KARL. Oh! It's magnificent! Stenson doesn't make trifling offers. He's like the tempter. He takes you to a high mountain and says: "I can give all this to you, if you will only give yourself to me." He offers me neither more, nor less, than the position of operating the factory at Lido.

ERNEST AND MDM TORELL. Papa's position?

MDM TORELL. And with the same salary?

KARL. He doesn't give as large a salary at first, but he promises to in the future.

MDM TORELL. Karl, dear, it's not possible to imagine greater happiness!

KARL. Happiness?! Can you really be saying this, Mama? Urging me to totally betray the memory of our father and forget all his work and ideas...

MDM FREDGOLM. No, I must speak my mind. This is really going too far. If I could even imagine my Fritz – who is, in fact, the same age as Karl – receiving such an offer! For the time being, his father, thank God, has the means to support us! But, though I never interfere with Fredgolm's business, we have a lot of children. I say we can't be making sacrifices for those who are unwilling to help themselves. (She leaves.)

MDM TORELL (crying). Karl, really – how can you make the decision to ruin us, now? Why have you taken it upon yourself to refuse help? We'll drive away all our friends!

KARL. And what do you say to all this, Ernest?

ERNEST. I say, there's nothing to talk about. How could you agree to such a promise? We'll find another way, as always.

KARL. Yes, but it's impossible to dismiss this so easily. It's, truly, a great offer. Stenson even suggested giving you the position of agent for the factory.

ERNEST. Thanks, but that offer doesn't tempt me. I prefer to find a job here, where I can live at home and not leave Mother.

KARL. You imagine that you'll be able to choose! We're stranded and you're ready to desert.

ERNEST. I can see that you're ready to accept. But it would be awful – to change so much...

KARL. It's not a question of changing. It's a question of receiving earnings and gradually building capital, so that our hands will no longer be tied. Do you really think I plan to work for Stenson for the rest of my life?

ERNEST. So, you're considering accepting this position?

KARL. I haven't told him so. On the contrary, I'm revolted by the idea. But I don't see how you can dismiss the idea so carelessly. This business is too serious. Remember, too, that, unless we do this, we'll be forced to sell all the machines and then how would we be able to continue the work?

(A servant enters with a letter.)

KARL. From Baron Yullenyelem. *(Opens it and quickly scans it, muttering in a low voice.)* Nobody can sincerely sympathize with how profoundly I mourn... His extensive scientific work has left behind significant economic difficulties, and I deeply regret that these difficulties have placed such a heavy burden on your young shoulders. You had such brilliant, courageous hopes for the future. *(Crumples the letter, but then reads further.)* Money, due to me from your father... *(Stops.)* To him, also! And about this, also, not a word in the accounts! It's awful that Papa kept the books so inaccurately!

MDM TORELL. Karl, don't condemn your dead father.

KARL. I don't condemn him at all, but, still, this is awful! I have taken all this on, not knowing what I would find. How much does Papa owe Yullenyelem?

MDM TORELL. I really don't know. It was just lately, when Papa had started to be sick, that he wished to accelerate...

KARL (*looking in the letter*). Here's the promissory note. Cancelled! 10,000 krona. He throws it in my face as a bribe! As if he's going to bribe me! Mister Baron, you are a subtle person. You may not name conditions, as Stenson does, but you hope to obligate me even more completely. Well, you were mistaken in your calculations. (*Sits down and writes, reading aloud.*) Mister Baron, accept my sincere gratitude. I have the honor to notify you that I have accepted the position held before by my father and I hope within the next few years to pay the debt to you, which he has left behind, for which I recognize myself as much responsible as if it were my own debt. Herewith, I deliver a new note in place of the former one. (*Rises.*) So! Now, at least, I'm free.

MDM TORELL. Yes, God will bless you, Karl! You're a good son. I always knew it.

(*She leans on his chest.*)

KARL (*pushing her aside*). You don't understand. That's not why I did it. I don't deserve your gratitude, Mother.

(*He runs out of the room.*)

ACT TWO

A room in Gerrgamra with doors open to a balcony located in the back part of the stage with stairs to a garden. To the left a window and a door. To the right two doors to other rooms. Baron Yullenyelem lies in an armchair near the balcony doors, constantly looking at the garden. Aunt Amelia sits near him and reads the newspaper aloud.

BARON YULLENYELEM (*interrupting*). Excuse me, please, but look over there! Isn't that Alisa's hat there in the distance by the boardwalk?

AUNT AMELIA (*rises and goes to the balcony*). Yes, she's been out on the sea again. She's just pulling in the boat.

BARON YULLENYELEM. Is she alone?

AUNT AMELIA. Yes.

BARON YULLENYELEM. But where's Yalmar?

AUNT AMELIA. Well, Jacob, my dear. The truth is, Alisa's the one to blame for this. Why does she always behave so strangely? She's forever lying in the boat and floating on the waves or climbing on the rocks to the very top of the falls and sitting there for hours.

BARON YULLENYELEM. I wish she were sitting up there today. Now, we have to prepare her for his arrival.

AUNT AMELIA. Well, I'll tell her in a way that manages to stir up her feminine pride so strongly that she will not even want to make an appearance.

BARON YULLENYELEM. Her feminine pride? What do you plan to say?

AUNT AMELIA. Not once the whole summer has he visited us. And now he comes only on business.

BARON YULLENYELEM. But I had forbidden him.

AUNT AMELIA. Yes, but Alisa doesn't know that. I assure you, she's very offended by his inattention. But you men never see anything. You probably haven't noticed how she's ceased to be interested in anything, lately – how she's become indifferent to everyday affairs.

(*Alisa enters from the balcony in a straw hat and print dress.*)

BARON YULLENYELEM (*holds out his hand to her*). Well, here's my lovely girl, at last.

ALISA (*with vivacity runs up to him*). And how are you feeling today, Papa?

BARON YULLENYELEM. If only my lovely little girl were more cheerful.

ALISA. Cheerful! Of course, I'm cheerful. Now, I'll take Aunt's place and read to you, Papa.

(*Goes to the door on the right and takes off her hat and gloves.*)

BARON YULLENYELEM (*quickly whispers to Aunt Amelia*). Leave us alone. I'd like to speak with her myself.

AUNT AMELIA (*also whispering*). Better it were I. Men aren't able to be so delicate.

BARON YULLENYELEM (*to Aunt Amelia, in a faltering voice, having noticed Alisa approaching*). I ask you. I ask you.

AUNT AMELIA (*in an offended tone*). Well, as you wish! (*rising and leaving*) It seems to me that after he has dealt with you so impudently...

ALISA. What are you talking about?

BARON YULLENYELEM. I don't understand at all what our darling Amelia...

AUNT AMELIA. Well, excuse me, Jacob, but I do not approve of your usual manner of always side-stepping delicate questions. I believe it's much better to tell the truth.

ALISA (*fitfully*). What truth? What do you mean? Why are you torturing me?

BARON YULLENYELEM (*with fervor*). Well, Amelia, if you refuse to allow me... Please, do me a favor and speak to her yourself. But, I am her father, and I think...

AUNT AMELIA. Please! I do not deserve to be reprimanded. God only knows how I've always tried to be like a mother to her.

BARON YULLENYELEM. Yes, unquestionably. Who denies it? But, now, my heart palpitations have begun again.

ALISA. When will you stop tormenting me? What are you talking about?

BARON YULLENYELEM. You know that I wrote…

AUNT AMELIA (*simultaneously*). You know that Jacob wrote…

(*Each looks with reproach at the other and stops.*)

BARON YULLENYELEM. Well, you tell!

AUNT AMELIA. No, you tell!

ALISA. It's simply heartless and cruel to torture me so!

BARON YULLENYELEM. Wait! Hush! Do I hear the clatter of a carriage?

AUNT AMELIA. (*runs to the window at the left*). They're here!

BARON YULLENYELEM. And, I wasn't able to prepare her!

ALISA. Who is it? (*She runs to the window and comes back slowly.*) Why is he here?

BARON YULLENYELEM. I really don't know. This morning I received a note from him asking whether I could receive him on business. You remember that I offered to have him disregard his father's debt. He refused the offer with extreme arrogance, and now I expect he repents and is coming to ask me to delay the debt. Amelia, would you please do me a favor and go receive them?

AUNT AMELIA. Yes, but I think Alisa should come along. His sister is with him.

ALISA. Oh! But first I need to… I've torn my dress on the boat.

(*She runs to the first door on the right.
Aunt Amelia leaves to the left.*)

BARON YULLENYELEM (*shouts after Alisa*). Alisa! I'll send his sister to you. You don't need to come to us. This is business, don't you see…

ALISA. And how do you think to settle it, Papa?

BARON YULLENYELEM. Oh! You can rest easy, my child. I'll not be harsh.

(*Alisa leaves.*)

(Karl and Paula accompanied by Aunt Amelia enter from the left.)

BARON YULLENYELEM *(makes some steps towards Karl and bows)*. Excuse me that you find me ill. *(He offers Karl a chair.)* And you, Miss Paula, will, perhaps, pass to my daughter's room. Amelia, would you please do me the favor of escorting her?

BARON YULLENYELEM *(again takes a reclining position in an armchair)*. It's been a long time, now, since I've had the honor. *(He suddenly extends to him a hand.)* Allow me to tell to you how deeply I respect you for your restraint.

KARL *(coldly)*. I've dared to come here today for the sole purpose of discussing business with you.

BARON YULLENYELEM. Yes? Yes, I understand, certainly, that your refusal of the offer I had proposed may have been too hasty, and that, now, perhaps, you – in which case I renew my offer.

KARL. That's not the issue. As I already had the honor to write to you, I have received a position operating the factory at Lido – my father's position.

BARON YULLENYELEM. Yes, a very responsible post for such a young man.

KARL. I understand this. For this reason, I feel duty-bound to make every effort to be worthy of the trust the company has bestowed on me. The problem is, the more I discover about the system, the more I become convinced that to run the business the way it's been run, up to now, is impossible. It will be necessary to develop the manufacturing on a much wider scale to make it profitable.

BARON YULLENYELEM *(indifferently)*. I think you are probably right.

KARL. To accomplish this, however, there's a rather important impediment. The quantity of water to which we have access is insufficient.

BARON YULLENYELEM. The problem should be helped by your father's invention. Are you not continuing to work on it?

KARL. For the time being, I may not be working on it. But there's another, simpler way to acquire the necessary water force.

BARON YULLENYELEM. Is there something I can do to help?

KARL. Yes. It rather closely concerns you, Mister Baron. The owners of Gerrgamra have built dams on your land for as long as one can remember.

BARON YULLENYELEM (*half standing*). Well?

KARL. Owing to this, a significant amount of water is distributed to you. The company intends, through the court, to force you to remove these dams.

BARON YULLENYELEM (*jumping up*). So, in other words, you want to ruin the Gerrgamra factory. Because, as I see it, without these dams we can't hope...

KARL. I know this. Therefore, I have boldly taken the measure of coming to you, Mister Baron, with the following offer: Lido and Gerrgamra cannot long exist as two competing factories. It's absolutely impossible. All the calculations I've made have convinced me of it. Their futures can only be assured under one condition – that they form an alliance, so that the interests of both factories will have merged and will have become common interests.

BARON YULLENYELEM. You can spare me this offer! Mister Stenson is not the sort of person with whom I would agree to enter into a merger.

KARL. As far as I know, Stenson has never done anything dishonest.

BARON YULLENYELEM. Neither would I say that he had. But our principles are too different. For me, the most important thing is the welfare of my workers. They have been in my service since childhood, and their parents and ancestors served my ancestors, and there was always a special view of our family name in this business. The factory was always considered from the point of view of what benefit it brought to the working people, instead of from the point of view of income provided by them.

KARL. Excuse me, but, if I am not mistaken, there is yet another concern demanding support of the factory – one more important to Mister Baron – the matter of the estate.

BARON YULLENYELEM. Ah! So this, too! Be assured, I shall not concede that my daughter has lost the estate. I understand your tactics, perfectly – to force me to close the factory. By so doing, you would take away my daughter's birthright, so that I would have no reason to refuse you her hand.

KARL. Mister Baron, such accusations regarding me are not worthy of you. I have no ulterior motives. I am here only to pursue the interests of the company I serve.

BARON YULLENYELEM. But, how does it happen that the interests of the company did not demand this until now – and not during the lifetime of your father?

KARL. Because my father didn't understand the need.

BARON YULLENYELEM. Because your father was a person of noble spirit, sir. And if you had even a spark of respect for his memory…

KARL. That, in fact, is my only concern. No one has the right to confront me regarding the respect I'm obliged to pay to my father's memory.

BARON YULLENYELEM. But this recourse will not be possible for you. Even if you were to win the suit, I would not close the factory. My shoulders can bear a greater load than this. And, before I would leave the people without work, I would be willing to personally sustain any losses.

KARL. This is all very well, but if the factory constantly incurs losses, the day will come when…

BARON YULLENYELEM. Never, kind sir! The day when you expel my impoverished daughter from her rightful home to then play the role of her benefactor will never come.

KARL. Mister Baron, I can suffer much from you, a former friend to my father and a man whose daughter… but you go too far! I have come here to offer you a merger, because this is the only means by which I can rescue you.

BARON YULLENYELEM. Rescue me! Ha! Ha! Excuse me, but this is just too outrageous! *(He shouts.)* Alisa!

ALISA *(runs in)*. Do you feel faint, Papa?

BARON YULLENYELEM *(falls to the armchair and speaks in a weak voice)*. Send for the doctor, my child. I feel a new attack beginning.

(Aunt Amelia and Paula enter. Aunt Amelia runs to the left.)

ALISA (*kneels near her father*). Papa, Papa, how did this happen to you? *(to Karl)* What have you done?

BARON YULLENYELEM. Help me to leave. Get me to my room. *(He rises. Karl wants to help him, together with Alisa, but he pushes him away.)* If this was just your intention, Mister Karl, it was, obviously, possible. The doctor warned me of the chance of another attack.

(*Alisa helps him to leave through the other door on the right.*)

PAULA. What does this mean?

KARL. He's willing to give up everything, as long as he keeps Alisa. What did she say to you?

PAULA. Almost nothing. She's offended that you never came to see them.

(*Yalmar slowly enters from the garden with a cigar in his mouth. On seeing Paula, he quickly tosses the cigar and hat and goes towards her.*)

YALMAR No, really? Such a rare guest! And how timely! Come quickly with me to the concert hall. I've just received some new music.

KARL. You must not know that Baron Yullenyelem is ill.

YALMAR. Yes, he's been ill for a long time. Unless this is something special?

PAULA. He's had an attack. Alisa is with him.

YALMAR. Well, that's all right, if Alisa's there. Such attacks ordinarily pass quickly. So, will you go with me?

PAULA. Oh! With pleasure. *(to Karl)* What do you think? Is it all right?

KARL (*absent-mindedly looking at the door through which Alisa has left*). Do as you wish.

PAULA (*to Yalmar, who is heading for a door on the left*). I almost never play now. I sit at the writing desk all the time and calculate. I calculate, endlessly.

YALMAR. What an unsuitable occupation for you!

PAULA. We need to work with all our might now, so that we can begin to earn a little more money.

YALMAR (*still heading left*). Hush, Hush! Money! Such words from your lips.

ALISA (*coming out on the right, to Yalmar*). Where are you going? Papa's sick.

YALMAR. Should I go to him?

ALISA. Better wait a while. He seems to be falling asleep. *(Yalmar and Paula leave. Alisa, addressing Karl.)* How could you? How could you?

KARL. Don't condemn me, not having heard.

ALISA. Whatever it was – Papa could flare up, could tell you troubles, speak nonsense, but no matter what, no matter what – how could you dare?!

KARL. The decision at hand was too important. I had to discuss it with him.

ALISA. So, you really want to go against us?

KARL. Your father has told me decisively that, even if we should do damage to his factory, he will always manage to support his workers. For us, it's extremely necessary to expand our business. For me not to do so is to neglect my duties to the company now expecting me to work on its behalf.

ALISA. I could suggest an alternative.

KARL. Tell me what to do.

ALISA. No. If your own feelings haven't prompted you...

KARL. I understand what you want to say – that I should leave the position I just accepted. Isn't that so? I'm actually glad to have the opportunity to talk with you about it. If, before accepting this position, I had learned about the possibility of a collision between the interests of Lido and Gerrgamra, I would, most certainly, not have agreed to take the offer. But, now, within two months of accepting the position, what reason could I give for such strange behavior. To take that action is impossible, believe me. Anyone who would hope to make his way in the world and to create for himself a position could never choose to compromise himself through such an act. It's impossible.

ALISA. Certainly, when the first and main purpose in life is to create for yourself a position, as you have said…

KARL (*drawing nearer to her*). And for whom do you think I… Alisa, don't misconstrue my words in such an awful way. You know all about what I cannot say to you – what I do not dare mention.

ALISA. I know absolutely nothing about anything. For two whole months you haven't once shown yourself.

KARL. Don't subject me to such a harsh test. I have no right to speak to you now. I'm forced to remain silent.

ALISA (*with suppressed sobbing*). I, certainly, wouldn't begin to tempt you to neglect your duties to your company.

KARL. Alisa! Do you, truly, not trust me?

ALISA. Trust you! Trust you! What is there to trust? For two whole months you haven't even attempted to see me, and now, when at last, you've come…

KARL. I wanted to act fairly concerning your father.

ALISA. Concerning my father – whom you have, perhaps, killed.

(*She bursts out sobbing, but bites her lip and stamps
her foot on the floor to repress herself.*)

KARL. You're being unfair. I'm not guilty of anything. If I have not been coming here, it's only because I believed it would be dishonorable on my part to join your life to the life of a man as financially insecure as I am. I wanted to build a future for us first.

ALISA. First! And how many years did you plan to wait? Until you should become rich and powerful enough to deign to honor me with your offer?

KARL. But, believe me, I have behaved in this way only because I love you so deeply.

ALISA. No, no, no! That's not true! If you loved me, you would have behaved completely differently. Never in your life would you have allowed my father to extract from you any promises. You would have come to me and would have told me: I have nothing to offer you. I am poor and in debt, but I love you and we can work together and, together, we can try to create for ourselves a future.

KARL (*coming nearer to her and seizing her with both hands*). Oh! Alisa, had I known your thoughts! But how could I dare to pull you away from the conditions in which you have grown up, to which you have become accustomed?

ALISA (*silent*). If you loved me – it would, still, not be too late.

KARL (*drawing her to himself*). Yes, surely, all is not yet lost?

ALISA. In these two long months that you left me alone, I have thought up the whole plan. I'll deliver to you the money necessary for your invention. Just relinquish this position.

KARL (*receding*). You, Alisa! How are you able?

ALISA. Wait, and I'll tell you. I have money that's mine by right of succession from my mother – too little for me to call myself wealthy, but it's enough, that…

KARL. And you would have me take that away from you?! How can you think…?

ALISA. Allow me to finish. I'm still a minor. But I have a very kind uncle on my mother's side. He has no children and he's very wealthy. I could borrow the money from him and repay him when I reach my majority – in less than two and a half years.

KARL. I do thank you, Alisa, for making such a noble and touching offer. I'll never forget…

ALISA. You see, everything can still be arranged! But you haven't responded to me, yet. You mustn't stay in the position at Stenson's.

KARL. But don't you see how impossible your offer is?

ALISA. Again, this intolerable word: *impossible*.

KARL. Anyone who would accept an offer based on such a relationship to the person he loves and to all her family would be unworthy of your respect – especially after what's taken place between your father and me. You're so noble that you don't understand all this. You yield your will to noble promptings and try to follow them, but a man has no right…

ALISA (*interrupting him*). And so, you insist that you love me, when everything else is more important to you than I am – when you're forever allowing all sorts of external influences to compromise our relationship. Why should I expect anything different from you? You've changed the memory of your father and ended the scientific activity. Why should you not abandon your feelings towards me, too?

KARL. No, Alisa, even from you I can't accept such words.

ALISA. Why should I continue to suffer, while you constantly push me away? This will be the last time!

(She escapes to the balcony, leans on a balustrade and bursts into tears. Karl follows her.)

ALISA. Just leave, leave! I'm asking you to leave!

(Karl, after a small hesitation, descends from the balcony and leaves through the garden gate. Yalmar and Paula return.)

PAULA. Karl has left?

ALISA. Yes, he's waiting for you near the carriage.

PAULA *(to Yalmar)*. Allow me to thank you, Baron, for giving me the pleasure. *(Extends to him both hands.)*

YALMAR. Please, come again – next time sooner. Why should we miss the pleasure of playing together? What do you say, Alisa?

(Alisa looks seriously at both, not answering.)

PAULA. Good-bye, dear. *(She kisses Alisa.)*

ALISA *(absently)*. Good-bye! *(Paula leaves on the left. Yalmar turns to follow her. Alisa makes a sign for him to remain.)* Yalmar! We should go to Papa now.

YALMAR *(coming back)*. Yes, if you think so.

(Paula leaves.)

ALISA. Do you love her, Yalmar?

YALMAR *(with surprise)*. I? Love? Such a high-sounding word! Certainly, I very much sympathize with her. I like her childlike naïveté, her great musical talent, but, for us men, such qualities don't hold as much importance as they do for you. And, besides, you know, I'm used to thinking of another person as my future wife. If she'll not be mine, I'll stay an old bachelor forever.

ALISA. Papa is very ill, Yalmar. You know that that would please him most of all.

YALMAR. So, you agree to it, Alisa?

ALISA. Yes. Now, I agree. *(Yalmar draws her to himself and wants a kiss.)*

ALISA *(pushes him away and speaks with suppressed, sobbing)*. That's not necessary. Let's go to Papa.

(She turns to leave through the door on the right. At this moment Karl and Paula come back and enter on the left. Alisa runs towards them.)

KARL *(silently approaches her)*. I'm not able to leave you like this. It can't be possible that everything I've dreamed of, everything I've lived for, for so many years, has now vanished in a flash. I ask you to talk with me alone for a few minutes.

YALMAR *(stepping forward)*. Forgive me, if I interrupt you, but my betrothed and I should go to our father now.

KARL. Betrothed?!

(Yalmar takes Alisa by the hand. They leave, Alisa with lowered head. Karl and Paula stand silent for several seconds.)

KARL *(angrily)*. Never again will we set foot in this house, Paula! Such unprecedented, such shocking change!

PAULA. *This* cannot be changed. Alisa loves *you*.

KARL. And Yalmar – *you*. Isn't that so? Didn't he just say something of that nature to you in the concert hall?

PAULA. He didn't actually say it, but it did seem to me…

KARL. It's disgraceful! How horribly they've deceived us! And, now, they're probably laughing at us upstairs – up there with the dying baron!

PAULA. How can you even imagine such a thing! Anyway, I, for one, am sure that they have not deceived us. And I'll always love both of them as strongly as I do now.

KARL. Only a common worm would allow itself to be so trampled. Let's get out of here! *(He leaves.)*

ACT THREE

The same room as in the first act, but in it are arranged simultaneously a drawing room and a dining room. The table is set for breakfast. Mdm Torell sits at the window and sews linens on a sewing machine. Paula runs in wearing an office worker's uniform, removes her apron and oversleeves, puts on a straw hat and shawl, and nods to her mother.

MDM TORELL. You're leaving now?

PAULA (*hurrying*). Yes, I'm already a little bit late. I didn't do very well today – several times I had to recheck my calculations.

MDM TORELL (*adjusting the machine*). It's like this with you every Saturday.

PAULA (*quickly turns around and comes nearer to her mother*). Has Karl said something?

MDM TORELL. Has he said anything to you?

PAULA. No.

MDM TORELL. So, clearly, I'm once again the scapegoat! I'm the one he'll blame.

PAULA. Well, what did he say?

MDM TORELL. In a word, he feels that lately you've stopped trying – as a matter of fact, all summer long – especially on Saturdays, when all you think of is how you'd like to break away for a walk. Karl says you're totally neglecting your duties.

PAULA. What should I do? Go back and take up the counting frame again? But how can I force him to wait so long?

MDM TORELL. Paula, my love, I'm afraid for you – afraid you'll be dismissed!

PAULA. So, Karl has said something?

MDM TORELL. Yes, he's becoming suspicious. Just think how he'd react if he were to find out how I've conspired with you.

PAULA (*sits down on a chair near her mother and nestles up to her*). Lovely, dear, sweet Mama, there's nothing bad going on. Even Alisa allows...

MDM TORELL. Yes, but does Alisa really know?

PAULA. That Yalmar has never been in love with her? Yes, certainly, she knows it. And she has at times said that I should be his wife.

MDM TORELL. Well, why did he not marry you in that case?

PAULA. He couldn't. The timing for both of us was unfortunate.

MDM TORELL. If he really loved you…

PAULA. Yes, if – but he doesn't love me – not deeply. Still, I mean more to him than anyone else.

MDM TORELL. And you can be content with this?

PAULA. I'm happy as long as I get to see him, frequently, and listen to him play.

MDM TORELL. But are you sure that Alisa accepts your frequent visits?

PAULA. She never says anything and always leaves us alone in the concert hall.

MDM TORELL. Well, you should, at least, put an end to these walks. Why should you be the one to go to him time and time again?

PAULA. But he can't come here to me because of Karl.

MDM TORELL. But, if you're meeting on a walk, he could come somewhere closer to here, instead of forcing you to go to all the way to Gerrgamra.

PAULA. It's so far to come here, and he doesn't like to walk.

MDM TORELL. He could ride.

PAULA. Yes, but then the driver…

MDM TORELL. He could drive himself.

PAULA. He says that it's boring.

MDM TORELL. Well, there you have it! There's nothing there – in the way of romance! He's not willing to endure even the slightest inconvenience for you. You, really, have no pride!

PAULA. Why should I worry about pride when I love him so!

MDM TORELL. Well, in my day, young ladies didn't behave this way – you can be sure of that. But, at any rate, don't you see that this is an inappropriate abuse of Alisa's trust?

PAULA. But we don't abuse it at all. We don't do anything wrong. We just walk together and play music together. *(Rushes, sobbing, to her mother's embrace.)* Oh! Mama, Mama! The truth is, this is all I live for. It's my one and only happiness in life!

MDM TORELL. I see and understand everything, my dear. That's why I don't have the heart to try to stop you. But if I could just, at least, discuss all this with Karl.

PAULA. That's impossible. He wouldn't understand anything. Since the time of his break with Alisa, he's become dry and cold. He would look at our relationship completely negatively – would not see, at all, what it is. While I, on the other hand, am so proud of it. I think it's much above and even better than ordinary married love. Understand, Mama, I would not like to be his wife at all. I wouldn't want to mar our love with every day prose. I want always to be to him only as much as I am now – nothing more.

MDM TORELL. Yes, you are probably right to think that Karl would consider all of this to be nonsense. But, he'll be so angry with me when he finds out. Besides, what should I tell him, since he already suspects something?

PAULA. Tell him just this – that his suspicions are completely unfounded. You can say this without qualm, Mama. Karl really has no concept of our true relationship. But it's time for me to leave. *(She kisses her.)*

MDM TORELL. Won't you at least have breakfast first, dear?

PAULA. No, thank you. I'm not hungry. *(She leaves.)*

MDM TORELL. And so it always is these days. You're absolutely damaging your health, my love.

KARL *(enters)*. Has she left?

MDM TORELL. Yes, she's left, as always, to take her Saturday walk.

KARL. Have you prepared steak for me, Mama?

(He sits down at the table, visibly tired.)

MDM TORELL. I'll get it now, dear.

(She goes into the kitchen and brings the steak.)

KARL *(eats with nervous haste)*. The breakfast wasn't even touched. So Paula ate nothing?

MDM TORELL. No, she didn't want to eat.

KARL. After four hours of work! With only one cup of coffee all morning! Where did she go?

MDM TORELL *(sits down to work)*. I don't know.

KARL. One of my technicians had business in Gerrgamra one of those days when Paula left to go for a walk. He saw her in the park alone with Yalmar.

MDM TORELL. Yes, she frequently goes to Alisa's.

KARL. With Yalmar – not with Alisa.

MDM TORELL. Yes, it's true that Alisa sometimes leaves them alone for a minute.

KARL *(rises)*. No, Mother. It's not necessary to be so blind. You need to talk to her, Mother.

MDM TORELL. You're not eating your steak, my dear?

KARL. No, because of this incident, I, too, have lost both sleep and appetite. Her calculations were incorrect again today.

MDM TORELL. Perhaps she's become somewhat overtired.

KARL. No, Mother. It's not possible to be so blind. Even, Martha...

MDM TORELL. Martha?

KARL. Yes, even she, so innocent and trustful, doesn't approve of Paula's conduct.

MDM TORELL. Has Martha spoken to you about it?

KARL. Yes – completely naïvely. This girl sees everything clearly.

MDM TORELL. You're apparently becoming interested in Martha.

KARL. Yes, I can't deny it. She's a lovely girl, in my opinion.

MDM TORELL. Lovely. Lovely, yes, but this is not what you need.

KARL. Why?

MDM TORELL. You always liked cleverer, more advanced...

KARL. Clever, advanced – bah! Driven, to a state of ecstasy in which it is impossible to live! No, better simple, uncomplicated natures, if it is necessary to marry at all. And it is necessary to marry, sooner or later.

MDM TORELL. But Martha loves Ernest.

KARL. Are you sure of that, Mama?

MDM TORELL. Actually, recently, she seems quite interested in you. Nevertheless...

(The parlor maid enters with the postal bag.)

KARL *(opens it and runs through the addresses on the letters)*. A letter to Paula from Yalmar.

MDM TORELL. How do you know?

KARL. I know his handwriting. Why would he write to her by mail instead of simply having the letter sent over?

MDM TORELL. Maybe he has no one there to send.

KARL. He – the fellow who never hesitates to tear workers off a plough any time he gets a notion. No, he didn't want Alisa to know about this letter, so he didn't dare send anyone.

MDM TORELL. What right do you have to be suspicious, Karl? He has written to Paula many times about her music lessons.

KARL. Because it's clear that there's something in this letter of a more intimate nature. *(Opens another letter.)* Well, here are new troubles.

MDM TORELL. Lord, help us! What now?

KARL. Well, this is the last straw to my patience! He has done enough nonsense.

MDM TORELL. Who, for Goodness sake?

KARL. Ernest! He received an order for our factory three months ago and hasn't even forwarded it to us. Now, they've turned to someone else and the company has incurred a loss of a thousand kroner, not to mention the scandal. I simply can't keep him in this position any longer. He damages our business.

MDM TORELL. But he's your brother, Karl, after all. You should be at least a little bit indulgent towards him.

KARL. I can't be indulgent, Mama. One can't be lenient with another when one has to struggle so hard himself. We need capable people.

MDM TORELL. If you don't have any feelings for your poor brother, at least have pity on me and don't worry me so terribly! Don't cause me such grief.

KARL. Mama, dearest, don't ask this of me. The situation is bad enough. Maybe you don't realize how much I have to endure, as I work day and night to get out of the hole into which we've gotten ourselves. It's terribly difficult to regain trust once your name... And how can I allow my relatives with their negligence and carelessness to turn all my efforts into nothing. There's no greater obstacle for a person wishing to make his way forward than to surround himself with worthless relatives. Paula already does everything she can to compromise me, and, now, Ernest.

MDM TORELL. I always said that it wasn't a good idea to send Ernest far from home.

KARL. Yes, if you had your way, he'd be staying at home, hiding behind his mama's skirt.

MDM TORELL. Ernest requires the gentle care of relatives. Everyone is not as strong as you are. Not everyone is so harsh.

KARL. What do you want from me, Mama?

MDM TORELL (*in a begging voice*). Obtain for Ernest a position here at home, Karl.

KARL. What kind of position? Would it be better to make him the manager instead of me? Then I would go somewhere far away, since I'm so harsh.

MDM TORELL. Oh! Karl! Why must you interpret my words that way? That's not what I meant at all.

KARL. Yes, if the question were just about me alone, I would go, willingly. God knows how much more difficult my situation is with the whole family against me.

MDM TORELL. Oh! Karl, Karl! How can you say that?

ACT THREE

(Enter Stenson and Martha. Karl goes with eagerness towards them and kindly greets Martha.)

MARTHA. Is Paula at home?

KARL. No, she's just left. She loves long walks, especially on Saturdays, when the work in the office comes to an early end.

MARTHA. Oh, I'd love to go out for a walk. If only there were someone... Do you ever go for walks Mister Manager? We could all three go.

KARL. Unfortunately, I've had no time – but for the pleasure of accompanying you, Miss...

MARTHA. Well, then... should we go together sometime to gather mushrooms? I'm very good at selecting mushrooms.

STENSON. Be quiet now, little chatterbox. I need to discuss very serious affairs with Karl. You can go to Paula's room for the moment.

MARTHA. Why may I not remain here? I won't listen to you. I'll just sit down here and sew.

(She removes her coat and hat and sits down at the machine.)

STENSON. I've come to talk with you about Ernest.

KARL. I already know all about it.

STENSON. Well... then, I hope you will have no objection to my plans. You need to know what was said today at the board meeting: Torell is a most efficient fellow, but he has one flaw in that he wants to push forward his utterly worthless relatives.

KARL. You see, Mother!

STENSON. Yes, dear Mdm Torell. There's no use crying about it. Ernest has to be dismissed – and the sooner, the better. We'll send him to America.

MDM TORELL *(jumping up)*. And you have the heart to do this, Karl?

(Martha quickly rises from her place.)

STENSON. What have you said to her?

MARTHA. Nothing.

(Turns away and looks out the window.)

MDM TORELL. Martha will, probably, say a kind word for poor Ernest.

MARTHA. I? Why would I?

MDM TORELL. I always thought that you were interested in him.

MARTHA. Oh, not in the least. I was certainly not. Quite the opposite. I always said that I could only be interested in a person whom I could marry.

STENSON. Yes, my girl has always had outstanding principles. She never foolishly filled her head reading novels, as do many other girls these days. For that matter, I don't think Martha has read even one novel through to the end.

MARTHA. No, novels always seem so silly to me. What pleasure is it to read fiction?

(Karl, quietly seeking her approval, says something to Martha.)

STENSON (addressing Mdm Torell). She would be a better match for Karl, wouldn't she?

MDM TORELL. But, you know, it's too bad for Ernest!

STENSON. Ernest! Are you out of your mind, Mdm Torell? (He gives her a friendly nudge.) We'll give him a tidy sum of money for the road and we'll send him to America. Believe me, it's the best thing for him.

KARL (quietly to Martha). So you were never seriously attached to Ernest, Miss Martha?

MARTHA. Certainly not. Paula had always wanted me to be in love with her brother, and when Alisa…

KARL. Actually, Paula wanted…

MARTHA. Yes, Paula and Alisa, were always such dreamers, such…

KARL. Yes, unfortunately for me. No! Girls such as you are made for family happiness.

MARTHA. Now you've said too much, Mister Karl.

KARL. Too much?

MARTHA (moving away from him). I don't want to hear anything at all that you would not repeat out loud. I never do anything behind my father's back.

(Approaches her father and hides her head on his chest.)

STENSON. Why, what's the matter?

MARTHA *(whispering)*. It seems that he has proposed to me, Papa.

STENSON. Well, this pleases me. You will have an efficient husband who will have an outstanding career, if only he can avoid excessive sentimentality, and you *(extending a hand to Karl)* will acquire a nice, thrifty little bride – simple and artless.

(They approach his mother and embrace her.)

MDM TORELL. I wish you God's blessing, dear children. But how could you forget Ernest so quickly?

KARL. It seems that was just pure childishness. Martha didn't understand her own heart. *(Caresses her.)* She's not as precocious as her girlfriends. There's still so much innocent directness in her that her soul lights up at the sight of a kind, innocent face.

STENSON. Well! Now, ordinarily, we would leave this young couple alone to cozy up to one another. But I still have affairs I need to discuss with Karl.

MARTHA. It's all right. I'll go to another room and chat for a while with my new mother-in-law. *(She approaches Karl and leans towards him for a kiss on the lips.)* Now it's permissible, since Papa and my mother-in-law know about it.

KARL *(kissing her)*. Ah, you, sweet innocent!

(Martha and Mdm Torell leave.)

STENSON. You will not regret having me as your father-in-law. The young Torell couple will acquire Gerrgamra as a residence sometime in the future.

KARL *(shudders)*. What are you saying?

STENSON. Only that the Yullenyelem family will soon be finished. They have, once again, taken out a large loan with indemnity. In a few years, they will be bankrupt.

KARL. And all this because of a lack of water.

STENSON. Yes, but why were they so reckless as to bring the matter to court. It's clear that we will win. And, most likely, it would please you to live in the castle with Martha?

KARL (*sharply*). Never! Don't even speak about it! This will never happen! On no account would I agree to it!

STENSON. So, as I said, you are still subject to sentimentality. It's a pity. It's a pity!

MDM TORELL AND MARTHA (*run in, shouting and interrupting one another*). Karl! The Yullenyelem carriage has stopped in front of the house! It's Baron Yalmar!

STENSON. Really? What does he need? Is this the first time he's come to you?

KARL. It will be better if we talk with him alone.

STENSON. The servant is submissive. Now he's being sent on his way. Farewell.

MARTHA (*approaches Karl and stretches her lips for a kiss*). And do you love your little bride?

STENSON. Go on! Go, my girl! It's never right to hold a man up with trifles when he has business to do. Women need to learn this!

> (*They leave, accompanied by Mdm Torell.*
> *Yalmar enters and bows ceremoniously.*)

KARL (*responds*). Sit down, please.

YALMAR. I've come on behalf of my wife, actually. She's come up with an idea that's perhaps completely impractical and possibly impossible, but, nevertheless, I didn't want to refuse to honor her request. You once helped your father in his work on the invention on which he had already made much progress. My wife feels sure – on what basis I don't know – that you were once close to a solution to the problem.

KARL. Yes, I had thought so.

YALMAR. But that you, for economic reasons, were forced to discontinue all work.

KARL. Not forever – but, yes, for a while.

YALMAR. She says that, now, engineer Grotto – a man you know – is working on solving the same problem.

KARL. Yes, he has told me so, but he doesn't have the means to finish, either.

YALMAR. Well, my wife now has it in her head to risk all our personal fortune on the search for the solution to this problem.

KARL. Really! And she's going to offer money to engineer Grotto?

YALMAR. No, my wife trusts you more. She suggests you should manage the money, however much is necessary.

KARL. She suggested me! I don't understand. It's impossible, that she really suggested me.

YALMAR. Yes, the truth be told, she's still very much the same young girl who once suggested giving you money on loan for the same purpose. At that time, certainly, you could not have acted differently than you did. But, now, your position is completely different. First of all, this is not a loan. All the risk remains on our side.

KARL. But I, really, absolutely do not understand what could induce the baroness...

YALMAR. Both my wife and I are at heart aesthetics, and this mad competition is unbearable to us. Alisa has always dreamed of making improvements in the lives of workers, and she realizes that, if we continue on the present course, this will not be possible. She's enthusiastic concerning your invention, because, in her opinion, it could resolve all of the difficulties. But, certainly, I understand that a businessman such as yourself could have a different point of view. Perhaps, it's in your interest to completely wipe out the factory at Gerrgamra?

KARL. Not so! In fact, as you probably know, I once made a proposal to your deceased uncle which should convince you that, quite the opposite, it was my hope to rescue Gerrgamra.

YALMAR. Yes, I am aware of your inclination towards good deeds. Still, it was not my intention at all to ask a favor of you today. I only intended to deliver to you the prospect of a pleasant opportunity to finish the business to which you and your father have already devoted so much energy.

KARL. Yes, and it is a great opportunity – certainly. But your offer is so unexpected. I can't give you a firm answer immediately – due to my position concerning my – concerning director Stenson's – company. However, I'm very grateful to the baroness for her trust – deeply grateful. In a few days, I'll be honored to come to you personally to discuss everything.

YALMAR (*rises and looks back*). How is Miss Paula?

KARL. She's not at home at present.

YALMAR. Really? I sent her a few lines by mail. It was concerning our occupation with music.

KARL. Here's your letter. It was brought in after she left.

YALMAR. What a disappointment! This means she'll walk such a long way in vain.

KARL. Such a long way? Apparently, you, Baron, are better informed about where she's gone, than we. And…

YALMAR. Yes, we were going to practice our music. Didn't she tell you? But I had written to her that we couldn't play today, since I would be coming here.

KARL. In that case, I'll order a harness and send after her.

YALMAR. No, no. Don't. You won't find her.

KARL. Not find her? I'll send someone to ask about her.

YALMAR. No, actually, I don't know, whether she will go directly there. It's a damned misfortune that the mail came so late.

KARL (*rises and strikes his hand on the table with a sudden burst of anger*). And you dare to come here with this offer to enter into a merger with you when you are regularly in confidential communication with my sister?

YALMAR (*coldly*). I don't understand.

(*Paula enters.*)

KARL. Back already! You could not have gone as far as Gerrgamra and be back in such a short time.

PAULA. I? As far as Gerrgamra?

KARL. Baron Yullenyelem has just told to me that you were to play today with him at Gerrgamra.

PAULA. How so? When the baron is here! I met the forest warden and he told to me that the baron had come here.

KARL. Therefore, you also came back. But you were going to go to Gerrgamra with the baron. Is that not so?

PAULA. No.

KARL. Now, I demand that you open this letter in my presence and allow me to read it.

(Submits the letter to her. Paula looks at Yalmar anxiously.)

YALMAR. What does this mean? Do you allow your brother such authority over you, Miss Paula?

PAULA *(to Karl)*. I don't understand how you can demand...

KARL. You do understand. If you will not prove to me now, with the help of this letter, the unreality of the awful suspicions excited in me by your behavior and your inconsistent signals, I shall immediately suspend all negotiations with the baron. And the existence of the surname Yullenyelem depends on these negotiations. Or, do you agree that I am right?

PAULA. I assure you, Karl. *(She opens the envelope, removes the letter and hands it to him.)* Take it! My conscience is clean. I dare to show it to you without even reading it.

KARL *(reads)*. I cannot come today to our appointment in the park, because Alisa will not allow me a minute of rest until I have seen your brother on an important business matter. Be at home when I come and then go down to the chapel. I shall get out of the carriage at the crossroads.

YALMAR *(barely constraining himself during the reading, pulls the letter out of Karl's hands)*. No! Such indelicacy exceeds my patience! Confine yourself to business, sir, and do not put your rough hands on that which does not concern you and which you cannot understand.

KARL. I understand one thing: if such a letter from a married man to a young girl can serve as proof of the innocence of their relationship...

PAULA. I swear to you, Karl. You don't understand, but it is so.

KARL. Yes, it's true that I don't understand how you, so kind and truthful, were able to go this far.

PAULA. If you had not prevented me from accepting from Alisa the loan to go and study abroad, this never would have happened.

KARL. Is that what serves as an apology for your shocking behavior concerning your friend from childhood? From this day on you will not set foot in Gerrgamra. I shall not allow you to disgrace yourself to such a degree.

YALMAR. Actually, now, by your rough intervention, you have managed to make our relationship no longer possible. I have but one thing to do: to say goodbye to Paula.

PAULA (*screaming*). Yalmar! Karl, I beg you! I can't live, not seeing him.

KARL. So, this is what it's come to! Anyway, it's not my fault if you're so lacking in self-respect that you can humiliate yourself so horribly.

PAULA. It's not humiliating, Karl! It will kill me! Only permit me to see him – not as frequently, if you want, but at least, though rarely, to see him!

YALMAR. Don't ask of him any favors, Paula. I couldn't respect myself at all if, after this, we should... when you have a spy.

(Presses her hand and quickly leaves.)

PAULA (*rushes, crying, to a chair*). Oh! Karl How could you?!

(Mdm Torell enters.)

KARL. Mama, you knew about it?!

MDM TORELL. It? What's happened? Oh! My Lord! Just look at Paula! This means that everything is in the open?

KARL. And you knew about it? Both mother and sister deceived me! Nobody trusted me. Everything was done behind my back. Why do you treat me as though I were a tyrant?

MDM TORELL. Oh, Karl, don't be so suspicious.

KARL. Suspicious! You call me suspicious, when I see how everyone betrays the person who would struggle honestly for the truth! How did Alisa treat me when, out of respect, I refrained from intruding upon her family before I had achieved a secure starting position I could offer to her? She found a way for meanness, for arrogance, with some understanding of my passionate love for her – a way to irritate me by falling into the arms of another. When I used all my efforts to get a foothold on the business, which my father had almost ruined, she threw in my face the charge that I had changed his memory! And when, now, I try to rescue my brother and my sister from destruction, by not allowing them to engage in bad, weak-willed acts, they say that I'm cruel! You do everything to ensure that I remain lonely, but even alone I'll manage to achieve success – by not going back a hair on what I think is right.

MARTHA (*runs in*). I'm eager to know what the baron wanted, Karl. Oh! Here's Paula! Has Karl told you the news?

PAULA (*barely suppressing her tears*). What?

MARTHA. That we're engaged! You are crying?

PAULA (*rises and turns to Karl*). If I supposed myself to have the right to interfere with your affairs as you interfere in mine, I would tell you, now, that your actions are the actions of one who is either ill or weak-willed.

MARTHA. What is she trying to say?

PAULA. This is what! If Karl told you he loves you, he's deceived either himself or you. He's never loved anyone but Alisa.

KARL (*sharply interrupting her*). By what right?

PAULA. By the same right as you. And I find your actions to be a thousand times worse than mine.

(*Paula runs out of the room.*)

MARTHA. She hasn't congratulated me at all! It wasn't nice of her to arouse jealousy in me towards Alisa. (*tenderly to Karl*) But she can't possibly. I believe you.

KARL (*drawing her to himself*). Thank you. Yes, trust me! Always trust me and be frank and truthful. Don't look on me as on a tyrant who needs to be deceived behind his back, because then I could become that person in reality. Be honest concerning me. Above all be honest with me!

MARTHA (*laughing*). Do you hear that, Mama! Be honest! As though I have something to hide. I hate secrets. But you should also be honest with me. What did Paula mean when speaking about Alisa? It wasn't nice of her. Let's assume I'm not worried about it, but... *(mysteriously)* Just tell me one thing – whether you ever kissed her.

KARL. Kissed – Alisa?

MARTHA. Yes – on the lips?

KARL. No, never.

MARTHA. No? Well then, in that case, what was she talking about?!

KARL. What has gotten into your head? Perhaps you... How close were you with Ernest on the Eve of Ivan Kupala?

MARTHA. Aren't you ashamed to ask such a thing? *(Puts her hand to his mouth.)* Ick, Karl! How horrible that you could think...

KARL. Actually, I don't think this in the least.

MARTHA. In that case, as your reward, here's a kiss. *(Kisses him.)*

MDM TORELL. What a lovely innocent child!

KARL *(with pathos)*. And thank God, she is a child!

CHANGE OF SCENERY

A big octagonal concert hall at Gerrgamra. A grand piano is in the middle. There are small bookcases and busts of musicians on the walls. There is a dome-shaped ceiling, a polished floor and carved window apertures without curtains. In a corner is a low, deep armchair, covered in leather. In the back are two doors: one to a study, another to Yalmar's smoking room with beaver skins, draperies and upholstered furniture.

(Alisa is going between the two rooms looking first out one window and then out the other. Suddenly, she throws the window on the right open wide and puts her head out. The clatter of wheels is audible. She shouts.)

ALISA. You're back, at last! Hello! I'm up here. Come up! You weren't gone long! *(She runs out the door on the left and in a minute comes back with Yalmar.)* Well, well, well?!

YALMAR *(dressed in a white raincoat and a wide hat with a white umbrella in his hands)*. Let me take off my raincoat. *(Alisa wants to help him.)* No, for heaven's sake, not here. You see how much dust is on me and the grand piano is open. *(He goes to the door of the other room and stops at the threshold.)* Why is the grand piano open?

ALISA. It was open when you left.

YALMAR. And nobody thought to close it? *(He removes the raincoat in the other room then comes back and slowly, carefully closes the grand piano.)*

ALISA. Tell me, Yalmar! What did he say?

YALMAR *(occupied with the grand piano)*. I never in my life met such a Philistine.

ALISA. A Philistine! About whom are you talking?!

YALMAR *(during the conversation goes back and forth between the two rooms, lights a cigar and begins to smoke)*. About whom am I talking?! That's a strange question!

ALISA. About Karl?!

YALMAR. Yes, about Karl — about wonderful Karl. I have never met another such dry, prosy, insensible, unscrupulous, petty bourgeois in my life! Thank God!

ALISA. Didn't he want to?

YALMAR. Truly, one would have to be completely naïve to believe that such a person is capable of invention. He only brags and shows off – that's all. If you could have heard the shameless way he refused my offer!

ALISA. He refused?! I didn't expect this – not now.

YALMAR. You may imagine that the basis for his refusal is something romantic – such as fear of seeing you and falling in love with you again. That's not it at all. I have found him, on the contrary, to be in the position of a just engaged, very happy groom.

ALISA. A groom?! Who is it you're talking about?

YALMAR. About Karl! All this is about Karl. All this time I haven't mentioned anyone other than Karl. Why should I repeat his name each time? And so, Karl has just become engaged to your former schoolmate, Martha Stenson.

ALISA. Karl – engaged to Martha? It's impossible!

YALMAR. I realized it immediately from the mood all through the house. She left when I entered and came back when I left. But I received inside information from the gold master to whom I gave your bracelet to be repaired. He told me that she had just come into the shop with her father and had ordered wedding rings.

ALISA. Martha! Well, there's a suitable wife for him! Pretty, innocent little Martha! And now he's so happy and so in love that nothing else matters to him. How touching! Listen, Yalmar!

YALMAR. Well?

ALISA. We'll write to engineer Grotto.

YALMAR. For revenge?

ALISA. Certainly not. How could I possibly make him angry – he who's found such happiness in the embraces of Martha? What does any invention matter to him, now – even if it has turned the whole world upside down – provided he's left alone and free to coo with his Martha? But, all the same, I want the work on the invention to be finished. I want to prove to him that it could work. And when, at last, he wakes up from his love trance – because his honeymoon cannot be eternal – he'll see, to his surprise, that, while he slept, the world moved forward with great force.

YALMAR. I can see that my news has struck you to the heart, Alisa. I didn't think it was so serious.

ALISA. Serious! As if to the grave I shall not be in love with the person who is so charming, so irresistible that he has won the heart of such a girl as Martha!

YALMAR. And you sent me – with the feelings for him in your soul still so much alive – you sent me to him to suggest a renewed acquaintance.

ALISA. No, Yalmar. You needn't think that. You know how this unnatural competition has tormented me. Only for that reason...

YALMAR. Perhaps you didn't even realize it. But what am I to think, now, when I can see...?

ALISA (*coldly, rejecting him*). I would not have troubled you to go had I not known that you would, at least, be pleased to see Paula.

YALMAR. You say this with such sarcasm. You're jealous!

ALISA. Jealous? Oh no, I'm quite used to others being loved more. At school they always said that I was the most capable, but I always knew that it was a mockery of fate to bestow on me such brilliant powers just to make me better able to understand how much more I could have been if anyone had ever really loved me with their whole heart.

YALMAR (*coming nearer to her*). I'm not going to be seeing Paula, anymore.

ALISA. How could you decide this?

YALMAR. You mean, you know what she was to me?

ALISA. Do you think I'm blind in both eyes?

YALMAR. And you never said a word to try to prevent...

ALISA. I felt I didn't have the right. She was for you what I could never be.

YALMAR. But you, of course, suspected.

ALISA. That you were unfaithful? Oh! No! I never thought that you, and especially not Paula, were capable of being unfaithful.

YALMAR. But I hope that we didn't hurt you.

ALISA. Of course not. Of course not. It was such a pleasure to me – so very pleasant – to see how everyone was loved except me – to have around me sights and sounds full of inexpressible tenderness – and to feel myself standing outside of it all. I always expected that at any moment our little Jacob… *(Rising tears interrupt her voice.)*

YALMAR. Dear Alisa, have you really felt so alone?

ALISA. Alone! In fact, it seems as though I've lived only to lose, one by one, everyone who ever loved me. First Karl. For him, career, position, external splendor held more value than my love. Paula always said I was for her a second self, until she met you. Then, I became nothing to her. And you! When we made our vows, you assured me that you loved me as deeply as you are able to love. But, I had not had time to become a wife to you before you sought in someone else the happiness you never found in me. Even little Jacob – I assure you, I'm perfectly sincere – surprises me every time he stretches his arms out to me instead of to someone else.

YALMAR. And you kept all this inside, not saying anything to anyone about it.

ALISA. Rather than ask for love, only to be rejected.

YALMAR. But I always felt that you didn't really love me. This was the reason I was distant towards you.

ALISA. But I was not indifferent to you, Yalmar. Believe me, I always loved you. And I, so sincerely, so passionately, wished that you could be at least a little bit interested in me. I didn't want much. All I wanted was for there to be no one else who was closer to you than I was. I've dreamed of only one thing all my life – to be the most important person in the life of another.

YALMAR. You're the most important person in my life, now.

ALISA. Yes, I could be that for you. *(He sits down in the armchair. She kneels at his feet.)* Between us there are so many ties. There is so much connecting us! Remember our happy childhood? All our shared memories here at Gerrgamra have become part of who we are – dear to both of us. And our little Jacob? He could help us in so many ways to become one soul – to help us begin to live one shared life – couldn't he?

YALMAR. Yes. We're branches of the same tree. I often feel how much we have in common.

ALISA. You're right, Yalmar. We have so much, so much in common! Why shouldn't we be happy? Let me show you how I can be when I feel loved! You will be amazed! Your poor Alisa is not without resources. Even Martha is not as good as poor, despised Alisa can be, if one looks at her with loving eyes. Look at me! Really look at me! Am I good? Yes. When I feel loved, I am good – but only when I feel loved – not otherwise. Am I kind? Yes. When I feel loved – I am kind. In fact, I'm not egotistical – I'm not self-centered. I can be completely free from self-interest and merge all my thoughts with a person who is truly close to me.

YALMAR (*drawing her to himself*). Now, you're the Alisa I remember – the Alisa of my childhood. As a child you were able to charm everyone, to do with them whatever you wanted.

ALISA. You see, now! I have qualities others don't have. Allow me to act in my own way with the engineer and the invention. Be for me what I ask you to be and I will be for you everything you could want. Through me you've received Gerrgamra. Now, I will rescue it for you – and for our little Jacob.

YALMAR. Yes, this invention is actually beginning to interest me.

ALISA. You see! The invention is even beginning to interest you – so here we have something else we can share.

YALMAR (*a little embarrassed*). Actually, I've been thinking about it for a long time now. I am even inspired to compose a symphony about it. I would call it "The Waterfall."

ALISA (*rising from kneeling at Yalmar's feet*). Forever this – the one thing you never share with me.

YALMAR. Listen, Alisa. This won't do. If you're going to be jealous of my music, we'll never be one.

ALISA (*coming nearer him again*). No, I won't be jealous. Only tell me how it goes – allow me to get a sense of what you're imagining. A musical narrative – "The Waterfall" – our waterfall, of course! It's a great idea! First you would need to portray it in primitive times, flowing freely in a wide, rough stream from the region of eternal snows – from the glaciers.

YALMAR. That's it! That's just how I imagined it! How did you know?

ALISA. That's how I am with someone I love. I know everything. I grasp everything. I understand everything. I could even become musical to live in perfect harmony with a performing artist.

ALISA (*continuing*). So, then, the construction of the factory begins – dams are constructed and the waterfall is forced to flow through a certain channel.

YALMAR. Yes, of course, and a deafening grumble from the waterfall!

ALISA. And the scratch of a saw, sharply and disharmoniously rushing to join the powerful roar of the water.

YALMAR. A saw?? Yes, it seems right! You're a genius. (*He embraces her.*) What wonderful discords could be created on this theme!

ALISA. And then – the spirit, the spirit of the waterfall, takes revenge on the people – as the waterflow continuously decreases and decreases.

YALMAR. Yes – and at last it turns into a small, insignificant stream and work at the factory stops.

ALISA. But, here, the invention comes on the scene. By the power of the invention, the people seize the spirit of the waterfall, completely subordinate it, and force each drop to serve to their advantage.

YALMAR. Yes, but it doesn't end there! Soon, people will want to take advantage of the power of other water flowing throughout the land. Then, the resistance of our waterfall will end, as all the other streams merge into a powerful chorus, audible from everywhere – from the ends of the earth! Then, our waterfall will not want to be left behind. Its voice will be heard and will sound, at last, with increasing force, until its roar finally covers the entire orchestra with its high silvery tone. Wouldn't this be an excellent finale?

ALISA. Yes, excellent. (*She moves away by several steps and speaks to herself.*) And I will, as before, remain alone.

ACT FOUR

The same room as in the previous scene.

(Yalmar sits at a grand piano and plays.)

FOREMAN (*enters and stops at the door*). Mister Baron, please tell me what answer to give to the workers. More and more people are coming to ask whether it's true that production has stopped at the factory.

YALMAR. Haven't I told you never to come to the concert hall with business issues? How is it that you're so impudent as to dare to presume to walk in here?

FOREMAN. The baroness told me that if the baron has no time to come to the office then I must go to the baron.

YALMAR (*rises up*). You should know that I will never, ever, under any circumstances, allow you to disturb me here!

FOREMAN. Does this mean I'll not receive an answer?

YALMAR (*jumps up from his seat*). What is this damned persistence? Can you really not leave me in peace in my own private rooms?

FOREMAN. Well, if the baron doesn't wish to give me an answer, then I can't do anything about it if the best and most highly skilled of our workers should decide to leave us. Anyone who has any hope of landing a position somewhere else could not, in such uncertain times, choose to remain here.

YALMAR. Well, then, let everyone leave – and first of all you! Now, leave me alone!

(The foreman leaves. Yalmar sits down again to the grand piano and starts to play staccato, nervously. Alisa enters.)

YALMAR. Did you have any special purpose when you sent the foreman here, or is this just another display of that method of education to which you now subject me?

ALISA. Well, apparently, you're planning to sit here and play until the new owner of the property arrives asking you to remove your grand piano and clear this space for him?

YALMAR. Perhaps he also has a grand piano?

ALISA. He has? Who has?

YALMAR. He who waits. He will not wait for our ruin to get into our castle. He must be laughing quite a lot at our engineer and our invention. You're a clever, outstanding woman, Alisa, but you're not gifted in two areas: in musical composition and in business. *(He begins to play again.)* By the way, I wonder if you can guess what it is I'm playing?

ALISA. Yalmar! Don't torment me! Really! This is such an important issue.

YALMAR. I was absolutely right. You're completely lacking in musical understanding. Because, if you were able to understand that I'm playing my own funeral march, you might, after all...

ALISA. This is not right. It's beyond my endurance. At such a moment, when the matter at hand is the resolution of such an important question, you suddenly present a farce.

YALMAR. Excuse me, my friend. It seems I'm forever trapped. I suppose I'll never learn to understand what's important and what is not. And now, I have made the error of having such a high opinion of myself that I even imagined that my leaving this world might be a more important issue than the closing of a factory. Yes, it's a known fact – I am an incorrigible egoist and, what's more, I'm aesthetic. To me the idea is absolutely intolerable that I should be buried to the sound of banal church singing. But most of all, the idea is disgusting to me that the old head pastor should require for me a requiem mass. *(sternly)* Listen, Alisa, if you have any respect for my last will, you will not allow that disgusting old crow to stand and croak. That's all I need!

ALISA. Yalmar! I simply refuse to listen to you! As if it weren't enough to have such important issues that I need to be thinking about, to have to listen to you say such things...

YALMAR. Yes, you're right. Perhaps it's not so important after all. As long as the bullet is on track, all will be ended in a trice.

ALISA. Can't you be serious for just five minutes?

YALMAR. Very well, I'll try. Actually, isn't it strange that I can be so cheerful? You've had many complaints about me, Alisa, but you've never reproached me for excessive cheerfulness. No, for the moment, I'll not discuss my funeral.

ALISA. This is pure madness. What are you hinting at? This is so inexcusable that I don't even want an answer! What can we tell the workers about our decision to close the factory, or should we postpone this announcement for a while?

YALMAR. Why deprive the poor people of their illusions, if there is no emergency?

ALISA. How can you?! It's deceitful to detain them when they might be able to find jobs elsewhere.

YALMAR. Why make the effort to ask me when you know what you want to say?

ALISA. We could move to Stockholm. There I might obtain lessons and you could make money with your music, if you wanted to.

YALMAR. Don't count me in on your calculations.

ALISA. With such talent as you have...

YALMAR. It's also my misfortune. I have too much true talent to turn it into a craft. *(rises)* Rather than go around this world giving concerts and playing before crowded floors of idiots, I'd prefer to go to my room and send myself on to the next world.

ALISA. Now, there's a real man!

YALMAR. And you're a real woman! You have the unusual ability to drive me crazy, to irritate me until the blood rushes to my head. So, I'm afraid your comment has no effect on me! *(He grabs his head with both hands.)* Now, it will be at least two days before I calm down enough to be able to play.

> *(He departs to his office and locks the door behind him. Alisa walks back and forth excitedly. There is a knock at the door on the left.)*

ALISA. Come in!

> *(Paula hesitantly sticks her head through the door, but, at the sight of Alisa, quickly approaches her.)*

ALISA. Paula! Where did you come from?

PAULA. I'm sure you must be surprised to see me here. But your husband sent me a letter that has me terribly frightened.

ALISA. A letter! I didn't know you had continued to correspond.

PAULA. We haven't corresponded for at least a year now, and it was forbidden for me to come here. But now Karl, too, was disturbed when I showed him the letter.

ALISA. What's in this letter?

PAULA. Here, read it yourself.

ALISA (*takes the letter and reads*). Come to me one last time. I want you to learn a requiem march for me, which you should play as soon as it is required in Gerrgamra chapel. (*She breaks off reading.*) This is pure folly – just a farce! You don't know Yalmar as well as I do. Wait here! (*She approaches the office door and knocks.*) Yalmar!

YALMAR (*opens the door and appears in an elegant Persian dressing gown with a cigar in his mouth*). What does my sovereign order now? I had actually hoped that I might be allowed to smoke my cigar in peace – to calm my nerves a little bit. (*He notices Paula.*) Ah!

ALISA. I'm going to talk with the foreman.

PAULA. Alisa, I won't stay if it displeases you!

ALISA. Not at all. Why would it displease me?

(*She turns, eager to leave.*)

PAULA. Alisa! Karl is here with me.

ALISA. Karl – here!

PAULA. When he saw the letter, he brought me in a racing sled.

ALISA. Where is he?

PAULA. I left him at the front stairs. He was going to the stable.

ALISA. And his wife? She's with him, I suppose? It's such a convenient moment for the future buyers to have a good look around Gerrgamra.

PAULA. Alisa, how can you think…

ALISA. Please excuse me – I have work to do. (*She leaves.*)

YALMAR (*who during this exchange has had time to go to his office and change from his dressing gown into an ordinary suit, returns now and approaches Paula*). You have such a serious, anxious look, Miss Paula. This means you don't think all of this so extremely ridiculous.

PAULA. Ridiculous!

YALMAR. Yes, that's Alisa's opinion. She's ready to choke with laughter as soon as I mention the word funeral.

PAULA. Are you ill?

YALMAR. No, human nature, unfortunately, is not so wisely arranged that a person can fall ill at the necessary moment. Therefore, one has to help oneself.

PAULA. Are you unhappy?

YALMAR. Unhappy! Oh, no! Such strong feelings are not acceptable for us at Gerrgamra. We are engaged not in feelings, but in business. And, it seems, mine is ruined.

PAULA. Can I do anything?

YALMAR. Just what I asked you to do. Also, I just wanted to see you one last time. You have become older – things are not well with you. Have I, also, grown old?

PAULA (*not looking at him*). I don't know. It's no concern to me.

YALMAR. So! Perhaps you already… I kept waiting to hear about your engagement to some young man with a good position.

PAULA. You will never hear such a thing.

YALMAR. Never? (*He takes her suddenly by both hands and leads her to the window.*) Look at me, Paula!

(*Paula looks at him with eyes full of tears.*)

YALMAR. So, you still love me?

PAULA. I have never loved any other.

YALMAR. Even the way you say this is so incredibly stirring! What a difference between such guileless outpouring of sincere feeling and that nervous, unnatural tenderness… There is still an escape, Paula, for me, for us. We can escape from this sinking ship. It will be considered cowardice, I know, but, somehow, I feel that it would be the most courageous act of my life. Here, I can't make anything right. I can only increase the confusion and the disorder. But we two could begin life all over again – together!

PAULA. It's impossible.

YALMAR. Don't fail me now. Don't you understand that this is, for me, a desperate, extraordinary step? If I can envision throwing everything overboard, you can, too.

PAULA. Could we really be happy, forgetting our duties?

YALMAR. At such a moment, you dismiss me with a trite phrase.

PAULA. Doesn't the sense of duty…?

YALMAR. It's just a phrase – and what's more, one of the emptiest. Unless it's our duty to torment one another to death. Alisa and I do torment each other – and most of all when we try to perform what you call our duty – when we try to excite feelings – when we try to love each other.

PAULA (*is silent, with lowered eyes*). You have tried?

YALMAR. Yes, we made the most desperate attempts, but our natures in that regard are only slightly in harmony, so that the efforts we made just disgusted us. Actually, our duty seems to consist in continuing to spoil one another's lives.

PAULA. I don't know – Alisa was my best friend. I'm not able to behave badly towards her. How can I deprive her child of his father?

YALMAR. He can lose his father without your help, you know. Paula, I behaved like a coward when I didn't marry you out of fear of poverty. I know you could give me a bitter answer – you could say: You only come to me, now, because others have changed towards you! (*Paula makes a gesture of protest.*) But, no, you don't say this, because you don't have that mistrust, that bitterness that stems from unnatural affections. You love, because you are capable of a naïve trust that doesn't ask, doesn't argue. Trust me – in spite of all that speaks against me. Trust that, if I go on living, it can only be with you. Tell me, Paula, do you trust me? (*He takes her hands.*)

PAULA (*taking her hands away from him*). Oh, God! When you're speaking, everything seems so possible, so simple. But when I think about Alisa – about Mama, about Karl! How could I act in such a way towards them?

YALMAR (*turns away from her*). I see that life has changed you, already. You've become prudent. When you see before you a drowning man, you don't rush into the water to rescue him, but worry that your dress might be ruined. Yes, if life can so change even you – you, who before had been so intimate, so direct – it's not necessary to bother with it. So, farewell, Paula!

PAULA. What do you plan to do?

YALMAR. To leave, somehow, while hunting with a loaded gun.

PAULA. Yalmar!

(She tries to keep him there, but he pulls away, goes into the office and locks the door behind him. Karl enters.)

PAULA *(runs to Karl)*. Karl! You have to rescue him! It's because of me that you stopped supporting them. But I'll do anything you ask. I'll leave. I'll take a position as a governess – anything – only don't allow him to destroy himself because of me.

(Alisa enters, but at the sight of Karl, she stops suddenly and starts to leave.)

PAULA *(running up to her)*. Alisa, you should listen to Karl. He wants to help.

ALISA *(to Paula, while trying not to look at Karl)*. You poor thing. You were always too naïve and optimistic. Here is a person set on the special purpose of ruining me, and you say: Here is your savior.

KARL *(coming nearer to her)*. Surely, you don't think that. You can't think that I really wanted to ruin you!

PAULA. Can't you hear? He *never* wanted that. *(hesitantly)* Alisa, may I ask a favor?

ALISA. What do you want?

PAULA. It would just please me so much, if you have nothing against it…

ALISA. *(noticing Yalmar's absence)*. Yalmar has left?

PAULA. Yes, he's in there. But, if you have nothing against it, I would so much like to see little Jacob.

ALISA. Go to him if you want. *(She opens a door to the right.)* Over there – where there used to be a nursery.

(Paula leaves.)

KARL. Your husband brought all this on himself through poor management. Contrary to what you think, I tried to keep brakes on the activity at Lido for fear of damaging the interests of Gerrgamra. If not for this fear, I could have expanded my business considerably and made much more of it than I have.

ALISA. But, first and foremost, you considered your company.

KARL. Perhaps it never occurred to you that a man sometimes gives greater consideration to his personal values than to his business?

ALISA. Only, not you. In fact, you attained all the things you were striving for – position, influence…

KARL (*with bitterness*). It's true. I have realized everything I had dreamed of – all I had hoped for in my youth.

YALMAR (*leaves with a gun on his shoulder and on passing bows to Karl*). Where's Paula?

ALISA. She's gone to see little Jacob. Where are you going?

YALMAR (*carelessly*). To hunt a little.

ALISA. At this time of day?!

YALMAR. Yes, for some kinds of hunting any hour is suitable.

ALISA. Yalmar! What are you thinking? You're frightening me!

YALMAR (*in a casual tone*). My dear friend, you're being childish. Do you think I'm not able to aim a gun?

(*Alisa looks at him sharply.*)

YALMAR (*assumes an indifferent, cold look and calls*). Lady! (*A hunting dog, his pet, runs out from his office on being called.*) Please give my regards to Paula!

ALISA. What for?

YALMAR. What for? For politeness.

(*Leaves, accompanied by the dog.*)

ALISA (*stands motionless, looking after him*). Oh, it's nothing, of course!

KARL (*has all this time stood in deep thought, not really listening to the conversation*). Why didn't you trust me? Why did you act so hastily, putting him between us?

ALISA (*with anger*). Not that, again! It's bad enough that we acted the way we did, but now, standing here friend to friend, you still want to about talk about it! When we are dealing with such awful, such needlessly ruined lives, can you not understand that it's intolerable to speak...

KARL (*constraining himself*). Right. We need to talk about business. Gerrgamra does not have to be sold. I'll do anything to prevent to it. You could rent the factory to me and I would try...

ALISA. No. It's too late. Nothing will rescue us now. All the holdings of the manor are incorporated. We can only hope to pay off the debts if we sell everything.

KARL. And your son?

ALISA. Actually, I don't really want him to grow up to be the successor of a manor. I've seen enough of that life. It was Yalmar's misfortune that he never needed to work. I would prefer my son to be poor – even though this, also...

KARL. This, also, is a source of misery. If I hadn't had to struggle just to earn a living...

ALISA. You're bringing up the past, again. We'd better talk about whether your father-in-law is going to buy Gerrgamra.

KARL. To that I shall never agree.

ALISA. How can you prevent it? What if he buys Gerrgamra and gives it as a gift to his daughter?

KARL (*looks all around*). I shall never enter this house as the owner... if you leave it.

ALISA. You know perfectly well that all this is just talk. When you think of it, what reason could you give? A man can't take such actions. Recall your own words.

KARL. I'm telling you, it doesn't have to happen. Gerrgamra should be rescued – for you. When I think of what I've sacrificed! It was for the sake of keeping Gerrgamra for you that I stepped back, so as not to deprive you...

ALISA. Why must you incessantly speak about the past? Please, spare me! *(She stamps her foot in great agitation and bursts into tears.)* It's intolerable – to have sacrificed so much because of such trifles, and then to stand and argue about it. It's absolutely intolerable. It's like digging with both hands in a bleeding wound. It's unbearable, unbearable! *(Her words are interrupted by her sobbing. She runs to the window, opens both halves and leans out from it.)*

> *(Karl follows her and stops behind her, miserable, with his head lowered. Alisa, after a long silence, turns slowly and gives him a hand, then starts to cry, having covered her face with the other hand. He starts to kiss her hand. Just then, Martha enters wearing a cape and hat, dressed in a graceful and very fashionable suit.)*

ALISA *(pulls away her hand, wipes her eyes with a handkerchief and stiffly approaches Martha)*. Madame Torell! What a surprise!

MARTHA. You call me Madame Torell!

> *(Embraces Alisa, who tries to avoid the embrace.)*

KARL *(in a displeased tone)*. What are you doing here?

MARTHA. Papa needed to send for you and it seemed to me it would be a pleasant ride for me. You need to go home now, because that engineer...

KARL. Who?

MARTHA. ...the one who during the past year has been working on solving the problem with the invention. You remember him. Can you imagine? It seems it was possible to solve it.

KARL and ALISA. It was possible?

MARTHA. Yes. He has come to discuss it with you. He will be heading to Stockholm soon to show scientists his invention. I saw it. It's a small model like your father's, but with one difference. Papa says this model moves and his didn't.

KARL. That's the problem I never got the chance to work on and solve. Now, it appears my life's hope is ended.

MARTHA. But Papa says that you're not to be worried, because the engineer has grown poor and is completely in debt. Papa says that it's much better to be in the position of one who can buy all this with mere money.

KARL (*to Alisa*). Now, I have attained everything – is that not so? Everything to which I aspired!

MARTHA. Papa has charged me, also, to tell Alisa that, if it really is necessary to sell Gerrgamra, he will buy it – and he says that he will give good money for it – so that you, also, should not worry.

ALISA. Oh! Thanks. Perhaps, you'd like to look at our rooms, since you're already here.

MARTHA (*looking at Karl*). Well, I wouldn't want to disturb you, now. But this room is very beautiful and so original.

KARL (*angrily*). How can you be so insensitive? What are you doing here? I really did not want to bring you along with me. Leave, now. I'm not able it to endure this.

MARTHA. I will never understand Karl. He's always this way! When others are pleased, he always has to spoil their pleasure.

(*She goes out in tears.*)

ALISA. So, this is your idea of matrimonial bliss!

KARL. It would seem that you were a bit hasty in speaking about my good fortune.

PAULA (*comes running in*). Alisa! It's so terrible! Yalmar...

ALISA. What?

PAULA. A shooting accident, while hunting.

ALISA. While hunting! (*First there is silence. Then she speaks in an unnaturally quiet tone.*) He has died?

PAULA. Yes.

ALISA (*falls into a chair*). And I didn't believe him!

(*The following series of retorts are said very fast, one after another, with strong excitement.*)

PAULA. So, you knew it, when he left. Alisa!

ALISA. I really didn't believe him. I mocked him!

KARL. Did he really do this intentionally?

PAULA. Yes, intentionally. And I could have prevented him, had I been allowed to.

ALISA. Did he do this because of me? Don't you understand that it would have been a thousand times easier for me to lose him to you than this way?

PAULA. I was never allowed to be to him what I could have been. Oh! Karl, Karl! How could you take on such terrible responsibility?

ALISA. It's actually fair that we're all suffering together. What a wreck we've made of our lives. Any thought of happiness now is unbearable. I want to suffer. I long for suffering.

KARL. Only one thought is more unbearable.

(Alisa looks at him questioningly.)

KARL. The thought of how it might have been.

THE STRUGGLE FOR HAPPINESS

Two Parallel Dramas

Written by
S. V. Kovalevskaya and A. C. Leffler

Translated by
Sandra DeLozier Coleman

Part II

HOW IT MIGHT HAVE BEEN

A Drama in Five Acts

CAST OF CHARACTERS

Baron Yalmar Yullenyelem	Baron Kronstrem
Alisa, his wife	Miss Kronstrem, his daughter
Aunt Amelia, her aunt	Major Elfors
Mdm Torell	Mdm Elfors
Karl ⎫	Anne, their niece
Ernest ⎬ her children	Master Foreman
Paula ⎭	Erik ⎫
Manufacturer Stenson	Anders Gulte ⎬ workers
Martha, his daughter	Sven Karlson ⎭
Mdm Selen	Peter Ström, an old man
Lieutenant Selen, her son	Anna, the wife of Erik
Gerta, her daughter	Anderson, a shopkeeper
Count Schperling	Other gentlemen
Countess Schperling	Workers

The action takes place at Gerrgamra of the prologue of the first play.

ACT ONE

The same scenery as in the last prologue, but earlier in the afternoon. Alisa, Aunt Amelia, Mdm Selen and Miss Kronstrem sit on the terrace of the castle and embroider. Anne and Gerta, teenagers, play croquet with the Countess Schperling and Mdm Elfors. The young women play slowly and lazily, as if in a daze, and continuously glance into the depths of the scene.

MDM SELEN. The sun has moved over to our side. Wouldn't it be better to go inside now?

ALISA. I'll hold an umbrella over you, Aunt. *(She opens a huge, ungainly umbrella and holds it over Mdm Selen.)*

MDM SELEN. But you'll grow tired of sitting here with an umbrella in your hands.

ALISA *(suppressing a yawn)*. No, it's all the same to me, whether to hold an umbrella or to embroider.

MISS KRONSTREM *(an old maid, who wants to seem young)*. There's no way Alisa is going to be persuaded to leave the terrace until the workers have passed by returning from work.

ALISA. You're right. The only pleasure I have here at the factory is seeing the workers. When you never do any work yourself, it can at least seem as if you're participating in some way in the life and work of others, as long as you show an interest.

MDM ELFORS *(playing croquet)*. When the men finally return home, they're going to be unbearable.

ALISA. Oh, I hope they don't return before seven. The food won't be ready until then and, until they've eaten and had their fill of drink, they are intolerable.

MISS KRONSTREM. Men, in general, are intolerably selfish and materialistic. How can they spend whole days on such heartless pursuits as hunting?

AUNT AMELIA. And it's so cruel, in my opinion. It's simply the worst!

ALISA. Actually, it's perfectly understandable to me. They're forced to become cruel and soulless to kill the boredom of this idyllic country life.

MDM SELEN. It seems strange to me, Alisa, that you have so little esteem for country life, when you grew up right here in the village.

AUNT AMELIA. And she loved Gerrgamra when she was a girl! I remember how happy you were when you'd come home from boarding school and what plans you made.

ALISA. Yes. Well, that was when I still had wonderful dreams about life. As long as a person can continue to dream, a real life isn't necessary.

MDM SELEN. In any case, I wouldn't say that there's so little life at such a large factory – with such a mass of workers.

ALISA. Well, you see, that's just what bothers me! In the city, moving around exclusively in the company of people of our own class, it's only rarely that we think about the fact that not everyone lives as we do. But at a large factory, like ours, especially when you're the owner, you find yourself seeing things in a different way. Here, the difference between our lives and the lives of others seems to be unreasonably huge. For example, consider what an enormous gulf separates me from all the young people at the factory, even though I grew up among them and played with some of them when I was a child.

MDM SELEN and MISS KRONSTREM (*overlapping*). But this is absolutely natural. How would you want…?

ALISA. Sometimes, when I wander all alone here in the park, I listen to how they gather in the meadow and sing their favorite songs so cheerfully. I listen to them laughing and chatting and playing – all these young people. Once, I sat down on the ground and just cried over my noble loneliness. I was overwhelmed by an irresistible desire to just escape and join them.

MDM SELEN. Oh, my dear child! What an idea!

AUNT AMELIA. This is just childish chatter. She would never do such a thing!

ALISA. No, I wouldn't. But why wouldn't I? Because I know that as soon as I arrived, they would be embarrassed and confused. Just my being there would stop them from having fun. No, I could never have fun with them. I don't understand their jokes. I would have a negative effect on everything they did. The fact is, there's not just an external divide between us – there's an internal gap, as well. This is what troubles me the most! How different life could be if everyone had the same training and upbringing, if we had the same habits and customs, and if we could just be part of one great circle of friends.

AUNT AMELIA (*to Mdm Selen*). Think about what she's imagining!

MDM SELEN. What does Yalmar say about this?

ALISA. Yalmar! To him, it's all the same, as long as I don't actually *do such a thing*, as Aunt Amelia says.

MDM SELEN. I'm so sorry, dear.

ALISA. About what, Aunt?

MDM SELEN. That you and Yalmar so little understand each other.

ALISA. Oh! Well! So what? People never understand each other.

MDM ELFORS (*throwing down her mallet*). I don't feel like playing. *(She takes Countess Schperling by the hand.)* Let's go change clothes.

COUNTESS SCHPERLING. All right, but we just changed.

MDM ELFORS. Who cares? It will help us pass the time until the men arrive.

(They're laughing as they leave.)

MDM ELFORS (*as she moves away*). My pink dress, the one with the lace...

ALISA (*jumps up at the first sound of church bells*). The vesper bells! Now, they will come!

GRETA SELEN. Our men?

ALISA. No, I'm talking about the workers. And, yes, the gentlemen, too, only not those *you're* interested in, but the ones who work at the factory.

ANNE (*in a low voice*). Do you mean you're interested in *them*?

(She begins whispering with Gerta. Alisa goes out to the iron lattice with Miss Kronstrem. The young girls start a game of shuttlecock. The workers begin pouring out in groups. They all take off their hats at the sight of Alisa, who is friendly and nods to them.)

MDM SELEN (*to Aunt Amelia, quietly*). Alisa sometimes sounds bitter when she speaks of Yalmar. Aren't they happy in their marriage?

AUNT AMELIA. Oh, certainly, they're happy – in their own way. After all, there never was any romantic love between them. Their marriage is based on something much more solid – on a sense of duty – on respect for a memory...

MDM SELEN. Maybe so... But wasn't Alisa once attracted to someone else?

AUNT AMELIA. Yes, but it was just a young girl's fantasy.

MDM SELEN. Doesn't he work at the factory, now, this young man? I believe his name is Torell?

AUNT AMELIA. Yes, he works here as a mechanical engineer. But they almost never talk. He and Alisa have a strained relationship.

ALISA (*turning to a young worker*). Good evening, Erik. How are things at home?

ERIK (*submissively*). Our little girl is not well, so poor Anna has a hard time...

ALISA. I'm sorry! And Anna herself is so weak. I'll look in on her – in the morning. (*Erik leaves.*)

ALISA (*to Miss Kronstrem*). His wife, Anna, was my best friend when we were children. I absolutely loved her! I used to be just as friendly with her as I was later with Paula Torell. But now she doesn't treat me as a friend at all when I come to visit. Instead, I'm cast in the role of a kind-hearted benefactress making condescending conversation when I try to talk to her. Don't you see that there's something terribly wrong with this?

MISS KRONSTREM. But, Alisa, dear, how can it be otherwise?

ALISA. When I think about Paula, that my relationship with her could change as well, I realize that, in reality, it's essentially just as unnatural...

(Mdm Torell passes through the park.)

ALISA (*shouts to her*). Mdm Torell! Good evening! Won't you come and chat with us for a minute?

MDM TORELL (*stopping*). Thank you for the invitation, Baroness, but you have so many guests.

ALISA. Only relatives. They came here to hunt. Please come in! There's no one to be shy about.

MDM TORELL (*enters the terrace and says hello*). To tell the truth, all day my heart's been somewhere else.

ALISA. Oh, mine, too. I understand, perfectly. But she'll be here soon! (*She begins to make introductions.*) Miss Kronstrem – Mdm Torell. Mdm Torell – Mdm Selen.

(*Aunt Amelia whispers with Mdm Selen.*)

MDM SELEN (*indulgently*). It seems I had the pleasure of meeting Mdm Torell last year. You're the mother of Engineer Torell, aren't you? I remember he showed us an interesting electrical device he was working on.

MDM TORELL. Yes, yes! You're remembering my eldest son, Karl. He's finally finished his work on the invention you saw and is now completing the experiments. I also have a younger son, Ernest, and a daughter, Paula. We're looking for her to return this evening from a trip abroad.

MDM SELEN. Oh? He has completed it?

MDM TORELL. Yes, but he doesn't have enough money to put it into production. A few days ago, he wrote to the academy about it and, now, he's hoping, with God's help, to be awarded a grant.

ALISA (*to Mdm Selen*). Mdm Torell's daughter, Paula, was my best friend in school. You know her, Aunt. She's an outstanding musician and she's been studying abroad.

MDM TORELL. Yes, thanks to the kindness of Baron and Baroness Yullenyelem, she's been able to continue with her studies.

ALISA (*to Mdm Selen, who is giving her a questioning look*). Oh, it was only a small scholarship that we gave.

MDM TORELL. The baroness is so kind and so modest. (*She takes hold of Alisa's hand.*)

ALISA. And now I'm so happy about her coming back home! Oh, Mdm Torell, I hope you're ready for me to be tearing her away from you this summer.

MDM SELEN (*rising*). I think we had better go inside.

ALISA. Won't you come with us, Mdm Torell?

(*Mdm Selen and Aunt Amelia begin leaving.*)

MDM TORELL. No, thank you, dear Baroness. I always wander here, waiting for Karl to leave his office – to try to get him to walk with me – or else he would never get any fresh air. I'm so worried he'll ruin his health with his sedentary lifestyle.

ALISA (*her face takes on an alarmed expression at the thought of Karl being ill*). Yes, he often works through the night.

MDM TORELL. At night as well? I didn't know this. He never told me.

ALISA. Well, of course, I don't know any details. But someone said – one of the workers said – that the light in his room is on most nights.

ANNE (*throwing down the shuttlecock*). I'm tired of this game. Would you rather go for a boat ride?

GERTA. Only if Miss Kronstrem will go with us.

MISS KRONSTREM (*jumps up like a young girl*). That would be great fun! (*All three depart.*)

MDM TORELL. I shouldn't keep you from your guests any longer, Baroness.

ALISA. Oh, do sit with me just a little longer – at least until your son comes for you.

MDM TORELL. He'll never approach, if he sees me sitting here.

ALISA. True… But why does he never come to see us?

MDM TORELL. Oh, it's perfectly understandable. He's always mindful of having taken the baron's money without having produced anything – or, at any rate, he thinks you and the baron have doubts as to whether he will succeed with his work.

ALISA. I have no doubts, and my husband will be convinced, too, when he hears that Karl's been awarded a grant.

MDM TORELL. Oh, well – certainly! Then his situation will change completely! For now, though, you must understand that it's somewhat degrading to him to be considered in the same light as Anderson, the shopkeeper, or to be viewed as just one more worker among a mass of factory workers.

ALISA. I actually wouldn't think so. He always wanted, whole-heartedly, to be a man of and for the people. Otherwise, why did he feel the need to form a consumer cooperative and take so much care to improve the lives of the workers? This is important to me, too – so, I admire that in him.

MDM TORELL. Yes, he has many good thoughts – good convictions. But to love people is one thing – to live and work among them is another. It's easier to love people when you admire them from afar, as you do, for example, Baroness.

(Karl enters the park. Mdm Torell moves to go meet him.)

ALISA *(gestures for her to wait a moment)*. No, I want to try to persuade him to come to us. *(She approaches Karl.)* Your mother's at our house. Won't you come in for a minute and talk with us?

KARL *(raises his hat and continues on his way)*. I appreciate the invitation, but I really don't have time.

MDM TORELL *(goes to him)*. But surely, we'll have our walk, as usual, Karl? And don't you want to go with me to meet Paula?

KARL. No. I don't have time. *(He leaves, hurriedly.)*

MDM TORELL *(to Alisa)*. Goodbye, dear Baroness. Of course, Paula will come to see you as soon as she arrives.

(She hurries after Karl.)

ANDERSON *(the shopkeeper, enters the park, climbs the stairs to the terrace, gives a somewhat sideways bow to Alisa and speaks)*. May I have a few words with the baron?

ALISA. The baron hasn't returned home, yet. He's out hunting.

ANDERSON. That's fine... That is, I'd like to say... Well, it would be good if I could at least have the honor of speaking with you, Baroness. Would you care to hear what I have to say?

ALISA. Please... Go ahead. You do me a favor.

*(She sits down and extends her hand towards a chair,
but he continues to stand with his hat in his hand.)*

ANDERSON. I would like to talk with you about this cooperative. You may already have heard that Engineer Torell is letting me go.

ALISA. No. But why?

ANDERSON. That's precisely what I want to explain. I have, you see... You see, at times, I have expressed my own views on a few things – views that differ somewhat from those of Engineer Torell. But Torell has a problem in that he never listens when people try to explain their ideas to him. He won't listen to anyone. He just gives orders. *(He looks questioningly at Alisa. She makes a sign to him to continue.)* This is his manner, but I have a different manner. Sometimes I express my opinions, if I may put it that way. And, lately, I've been saying: Our business can't continue under our current plan with the store at Lido substantially reducing prices and extending credit. To compete, we'll have to follow suit and reduce our prices and also allow credit. Do you see my reasoning?

ALISA. But the whole idea of a consumer cooperative is to avoid selling on credit and to not lower prices in comparison with other merchants, but, instead, to deliver high quality products and then return a small dividend to the buyers at the end of the year.

ANDERSON. Exactly! Exactly! But here's the problem. A year has passed, already, and next week we're due to receive the dividends. What do you think will happen, Baroness, if no dividends are given out?

ALISA. But why would they not be given out?

ANDERSON. Yes, why not? That's exactly what Engineer Torell is going to have to explain. I don't want to burden the baroness with stories about certain facts that are indicated in regard to this question. Soon it will be clear where he used the money entrusted to him. But, here's what I respectfully ask of you, Baroness – that you would intercede for me with a word to Torell, telling him that he acted unjustly in dismissing me, since I have always tried to carry out company business in an exemplary manner.

ALISA. But you, obviously, acted against his orders. No, I will not interfere in this. I'm sure, if Engineer Torell has dismissed you, he has his reasons. And regarding your insinuations with respect to his honor, they do more to compromise you than him. Goodbye.

(She gets up and starts to leave.)

(Anderson, in a fury, pulls his hat over his eyes and departs swiftly. On his way he meets Karl and Mdm Torell, who are rapidly entering the park, but he goes past them without bowing.)

MDM TORELL *(shouts to Alisa)*. She's coming! We saw a carriage at the turn. *(She goes further into the depths and joins Karl.)*

ALISA. Ah! *(She runs after them.)*

ANDERSON *(to a factory employee, who was expecting him in the park)*. She became angry and now she's starting to defend him.

OTHER MAN. I always said as much. She's not impartial towards him. You should arrange to talk with the baron himself.

ANDERSON. Yes, definitely. I'll wait for him to return from the hunt. Let's sit down somewhere.

(They both depart.)

(Alisa, Karl and Mdm Torell return with Paula who is in her traveling suit. The three women speak excitedly.)

Overlapping
{
ALISA. Ah, how glad I am to see you!
PAULA. How happy I am to be home again!
MDM TORELL. My dear child, you're home!
}

(Paula embraces first one, then the other.)

ALISA *(to Mdm Torell and Karl)*. Now, you really must all stay and dine with us. We all want to see Paula, to enjoy her company, and she'll be torn if we don't stay together.

MDM TORELL. Thank you.

ALISA. And, it goes without saying that your younger son will join us, Mdm Torell.

MDM TORELL. Thank you very, very much for the invitation, dear Baroness *(turning to Karl who is standing at some distance)*. And, of course, you will dine with us, won't you, Karl, dear?

KARL. Thank you, but I've already had lunch.

ALISA. So, call this dinner, if you want.

KARL. Thank you, but I need to write a few letters.

ALISA. You must do this tonight? When Paula has just returned from being away for three years?

KARL *(dryly)*. I can finish and then talk to Paula later. *(He heads away.)*

MDM TORELL *(goes and speaks to him in the background with Alisa and Paula remaining in the foreground)*. But, Karl, dear, why won't you stay? It's so impolite.

KARL. You know very well that I'm uncomfortable here! How can we enjoy being with Paula, while she's here at the disposal of her benefactors? This is the way it always goes when you're obliged to people. It was so inconsiderate of them to take her away from us the minute she arrived.

(He leaves with his mother following him, talking and convincing.)

ALISA *(to Paula)*. I'm afraid your brother's not happy that I've kept you here tonight. This was, of course, thoughtless of me.

PAULA. Oh, of course not. Nothing could be nicer, than for us to be together. But where's your husband? I want so much to see him – to thank him for his kindness.

ALISA. He should arrive soon. He spends whole days hunting this time of the year.

PAULA. Oh? And, while he hunts, what do you do?

ALISA. I embroider. Or make conversation.

PAULA. And nothing else?

ALISA. Well, when the men return home, I play at the role of hostess at the table and am responsible for making sure the champagne is moderately cold.

PAULA. But, Alisa, is this a good life for you?

ALISA. No. This is not a life... After dinner, the men, having drunk as much as they need, and sometimes a little more, immediately retire to the billiards room and stay there until well after midnight.

PAULA. What do you do?

ALISA. I go to my room and lie down and read until the house becomes quiet. I don't even hear Yalmar when he goes to his room.

PAULA. Alone?

ALISA. Yes, alone. I don't see him 'til the next morning.

PAULA. But how did you develop this relationship?

ALISA. Develop? This is how it's always been. Yalmar and I have lived in Gerrgamra together all our lives in almost the same relationship we have now. Our marriage didn't change anything.

PAULA. And I had imagined that you were happy! During the most cheerful times I had in Leipzig, I found myself thinking: All this is nothing compared to that – *(quietly, turning her face away)* – to loving and being loved.

ALISA *(getting up)*. Yes... *(then shakes her head)*. Oh! Come on! It's really not so necessary. Life without love is quite possible!

PAULA. But you were making such grand plans for improvements for the people. Why don't you try to complete them?

ALISA. Alone? With no help and with no one who shares my goals? I can't.

PAULA. But why...? May I ask you a question, Alisa?

ALISA. Do me a favor.

PAULA. Why, since Karl is working to create a consumer cooperative with pension funds and other benefits for workers – why are you not participating?

ALISA. He doesn't want me involved. That's very clear. You saw for yourself how he treats me.

PAULA. But, what does it mean...? May I ask just one more question?

ALISA. What?

PAULA. I hope you won't be angry. But what happened between you and Karl? When I left, I was sure you and he...

ALISA (*stands up, somewhat shaken*). No, it's probably better not to ask. (*She goes and looks into the depths of the scene.*) It seems our hunters are finally returning.

PAULA (*also getting up*). I should go home and change.

ALISA. Go, but hurry back! (*She takes her hand and speaks seriously.*) You might want to ask your brother. I honestly don't know. (*She lets go and goes towards the men into the depths of the scene. Paula exits left.*)

MDM ELFORS and THE COUNTESS (*coming down the stairs*). At last, they've arrived!

(*Both the young girls and Miss Kronstrem run in from the right.*)

ANNE. Are they here already?

GERTA (*to Anne*). My braid's fallen down in back. Come help me.

(*They run inside. The others hurry towards the men, who are approaching from the depths of the scene in their hunting clothes with their dogs.*)

OVERLAPPING —

BARON KRONSTREM (*kissing his daughter on the forehead*). Well, how did you entertain yourselves today?

MDM ELFORS (*being kissed*). How was the hunting?

THE COUNT (*kissing the Countess's hand*). Can you imagine? We shot a huge moose!

LIEUTENANT SELEN (*approaches the stairs as the girls, returning from the castle, are coming down. He bows, gives Anne a colorful field bouquet and turns to Gerta.*) How is mother doing today? Yes! We downed a magnificent moose!

ALISA (*to Yalmar, who immediately drops into a chair and raises his obviously-tired feet*). You can't imagine how wonderful she is. It's going to be such fun having her here all summer!

YALMAR. Oh! Great! I can see it now. No rest in this house. This young girl, who's so enchanting, will, of course, come 'round to exchange courtesies by 10 o'clock in the morning. *(turning to the count)* She's just right for you, Schperling – you the music lover.

SCHPERLING. What! You actually recall that I'm a music lover. Usually, when I ask you to play, you ignore it.

YALMAR. When you hear this beautiful young girl play, I think her performance will make a much stronger impression on you than mine would.

SCHPERLING. Haven't you always wanted to find someone like this who could accompany you when you play your violin?

YALMAR. If she could accompany me hiding behind the curtains, somewhere... But in the same room, it would be hard.

ALISA *(to Schperling)*. Yalmar's only joking. He was an ardent admirer and maybe even a little bit in love with her.

YALMAR. Yes, that's what would make it difficult, now. She would, probably, imagine that I still care for her.

ELFORS. Oh! No! Married people aren't even a consideration. That accusation could only be directed at Selen.

YALMAR. But *you* forget about *your* wife, Schperling – a case in point.

OVERLAPPING.

> BARON KRONSTREM. Who is this girl you're talking about?
>
> SCHPERLING. She's the sister of that great inventor. Right? It seems this family has cost you a lot of money.

YALMAR. Not me. That's my wife's affair. Paula's her best friend. With regard to him, he came to us and we were interested. His invention had the potential to save us from the difficulties we fell into when Stenson acquired control of the dams.

ELFORS. And you think he'll finally manage to complete the work.

YALMAR. Probably not. I've already invested too much in this venture and I'm done.

ALISA. But the problem is solved in theory.

ELFORS. In theory! Ah-ha!

SCHPERLING. But, to put it to use, it lacks just one little thing. That's how it was with the fellow who thought he'd invented a perpetual motion machine. He was missing just a tiny part that would have made it go… *(indicating the motion with his finger)*

ALISA. But, before long, it will be obvious that he's right. He just sent a paper to the science academy in which he proves…

BARON KRONSTREM. Proofs are proofs, kind Baroness, but, for us factory owners, practical evidence is what's important. Let him produce a good storage battery and make it work. That will have a good deal more value for me than any academic recognition.

ALISA. He *will* put it into action – as soon as he gets the necessary funding.

YALMAR. If he's awarded the grant, then I might actually consider risking a few thousand more on this business.

ALISA. But, then he wouldn't need your help. He will only need help if he doesn't win the grant.

YALMAR. Thank you! Finally, we have the assurance we need that he is an outrageous swindler.

ALISA. But you know perfectly well that that can't be said of him.

YALMAR. On the contrary, my friend, there are many who say this. I just talked with Anderson. A pretty mess of stories is coming in.

SELEN. Yes, the consumer society – the modern way of building a work force.

YALMAR. And all for the purpose of climbing up the shoulders of the party.

SCHPERLING. A rather curious social democrat this gentleman is. When you look at the good-natured friendliness this pure aristocrat, Yalmar, always extends to the people and then compare it with the arrogance of this, so-called, *friend of the workers*…

ELFORS. Yes. It will be interesting to see how things turn out at the general assembly. I assure you, Baroness, what they're saying about him sounds very serious.

PAULA (*enters wearing a light, fashionable dress, rushes onto the terrace and approaches Yalmar, who slowly gets up from his chair*). I don't know how to thank you, Baron.

YALMAR. For heaven's sake, I'm not involved in this at all. I am very glad that it went so well. (*turning to the group*) Allow me to introduce to you a new artist, Miss Paula Torell. Count Schperling, who is a great connoisseur of music, Countess Schperling, Lieutenant Selen, also a great connoisseur, but not of music, and... Major Elfors.

PAULA. Have mercy! So many introductions at once! I'm not at all good at remembering names.

(*Yalmar sits down again in his place.*)

LIEUTENANT SELEN (*bowing and smiling*). What can I do to get you to remember my name?

SCHPERLING. Did you study at the Leipzig Conservatory?

PAULA. Yes. What a wonderful time it was! It seems to me that nothing in the world is more pleasant than our life there.

SELEN. So, you were very sorry to leave?

PAULA. No, I can't say that. To return home is also fine – to see all the old places – the faces of people who are dear to you. I think that there is nothing in the world more pleasant than returning home after such a long absence.

YALMAR (*considering his fingers*). Two.

PAULA. What?

YALMAR. Nothing!

SCHPERLING. And you performed in a concert. The newspaper wrote a lot about this, saying that you played your own compositions.

ALISA. Weren't you afraid when you played for the first time before a large crowd of strangers?

PAULA. Yes, at first, I was terribly afraid. But, then, I realized that I was understood! No, there is, truly, nothing in the world more pleasant than to be aware that everyone is being carried along with you – everyone – absolutely the entire hall.

YALMAR. Three!

PAULA. What's the matter?

YALMAR. You have just declared three things to be the most pleasant thing in the world. First, it was a life among comrades — that is, a life in perfect harmony with all manner of insignificant persons. Now, it's the approval of the public, two thirds of which consists of idiots with regard to their ability to appreciate music. And you also listed returning home after being away for a long time. Well, this is at least more understandable to me. After all the problems of being on the road, it is pleasant to lie down in one's own bed, to sit at one's own table. By the way, Alisa, when will dinner be ready? We wore ourselves out splendidly, today. What are you serving us?

ALISA. We just received the large batch of oysters you ordered from England.

YALMAR. Well, that's fine. *(turning to the men)* I specifically ordered English oysters. This is a very good grade — small, with a thin shell. They're much tastier than the Dutch. And you have chosen a Chablis, Alisa?

ALISA. No, champagne is prepared.

YALMAR. Champagne, my friends! No surprise — as women, even the smartest, have no sense in these matters. Who drinks anything other than Chablis with oysters? Oysters call for a good Chablis. I'll go down to the cellar myself. *(He exits.)*

ALISA. But, Paula, where are the others?

PAULA. Karl couldn't be persuaded, so Mama decided not to come. But Ernest will come.

ALISA. That's what I can't stand about your brother. It's petty on his part to always behave so distantly towards us. Can't you convince him that it's not kind — that it's not even polite?

(Aunt Amelia shows up at the stairs. Paula waits for her.)

BARON KRONSTREM. Ah! Here's Baroness Yullenyelem! A subtle hint, no doubt, that we should go and change our clothes.

THE REST OF THE MEN. Of course.

MAJOR ELFORS *(to his wife, quietly)*. Come. Get me a clean starched shirt.
(They walk out together.)

COUNTESS SCHPERLING. Why are you not going, Otto?

SCHPERLING. Will Miss Paula disappear before we've had time to return?

PAULA. Oh, no.

LIEUTENANT SELEN. Can this be true? In this village, a young woman is such a rare phenomenon... For her to disappear is just expected.

(All gradually enter the castle, except Alisa and Paula.)

ALISA. Well, Paula. You seem to be standing at your peak. Everywhere you turn, people are filled with delight and admiration.

PAULA. Do you envy me this?

ALISA. Yes... and why not? It must be nice to feel the force of one's personality having such a strong effect on others... Although, with me, it wouldn't be quite the same.

PAULA. What do you mean?

ALISA. To have power over another person... I've dreamt about it often – but only over one – over one and only one.

PAULA. There's Karl... I'll go talk to him again.

ALISA *(shudders)*. No, don't. He doesn't want to come on his own. As for myself, I have no desire to draw guests in by force. *(She goes into the castle.)*

PAULA *(descends into the park towards Karl)*. Where are you going?

KARL. Back to the office.

PAULA. Is this absolutely necessary? We could be having a pleasant evening here together. They're really so nice.

KARL. Yes, towards you, I can believe they are. You've justified their good deed to them. They can't be anything but proud of you. But, in their minds, I'm still just one freeloader among many others, who costs them a lot of money without any benefit. Their other freeloaders aren't invited to dinner.

PAULA. Actually, in the very near future, Gerrgamra will be obliged to you for its having been rescued.

KARL. Yes, I'm sure they will extend all manner of courtesy and attention to me when I finally finish my work. But then I will be even less interested in attending their social gatherings. If they would take me at my word...

PAULA. But Alisa believes in you.

KARL. The fact is – she doesn't. As if I care about any of the others. Even Alisa still needs concrete evidence. She doesn't trust my word. She's hurt me by this and I can never forgive her.

PAULA. But why do you say this?

KARL. Every day that passes that I don't have some new evidence, I read her doubts in her eyes, in the tone of her voice, in every hello. When the evening mail arrives, she immediately asks me if there's any news from the academy.

PAULA. Tell me what happened to you and Alisa, Karl. When I left, I was sure...

KARL. Do you really think that I could come in here as a simple machine designer, who had received a small loan of 15,000 krona out of grace, a debt which only added to the 10,000 that father already owed, and go up to old Baron Yullenyelem and ask for his daughter's hand, just to have him ridicule me?

PAULA. But if Alisa loved you?

KARL. She immediately gave her promise to Baron Yalmar.

PAULA. Yes, but only when you stepped back.

KARL. I don't see that I had a choice.

ERNEST (enters quickly). What do you say? Do I look presentable? It's downright elegant to be invited to the castle like family. We have our talented sister to thank for this honor. Do I look like a real gentleman? I chose the Johnson tailcoat – is that good? This pin from Berlin is so stylish. I wish Martha could see me!

PAULA. Ah, Martha. Are you still in love with each other?

ERNEST. Of course, although we still don't dare admit it. But when Karl gets his grant, then it will be different. All doors will be wide open to us. The great inventor and the brother who has been such a great help to him. The great pianist and her brother (turns around).

Your little brother is a chic young man, isn't he? *(He embraces Paula and begins to whirl around with her.)*

(Alisa and Yalmar approach and say hello to Ernest.)

YALMAR *(to Karl, who is turning to leave)*. Won't you stay for dinner, Mister Engineer...? Please.

KARL. Thank you. I just received a letter from the post office and should answer.

ALISA *(with animation)*. Could it be news from the academy?

KARL. No. *(He leaves.)*

(At this time, Schperling is shown on one side of the castle, Selen on the other, both with bouquets for Paula. Aunt Amelia emerges at the same time from the castle.)

PAULA *(laughing)*. How can I take them both at once? *(She takes a rose from Selen's bouquet.)* I'll put this one in my hair. *(She attaches it.)*

SCHPERLING. Don't make me envious. Or, else, Selen and I will shoot each other on the next hunt.

PAULA *(takes several flowers from Schperling's bouquet.)* No, there's no need. I'll fasten these to my chest. *(She attaches them.)*

YALMAR *(approaching, leans towards her)*. Let me help you. In view of the fact that it's my garden that's being stripped of roses for you, shouldn't I get *some* reward?

(Paula looks at him, blushing and slightly distracted.)

YALMAR *(quietly)*. We shared an earlier bond. You shouldn't show others a preference, now.

PAULA. You've changed so much, Baron.

YALMAR. Try me, first, and then judge. Have I have ever shown my true self in such a social setting?

PAULA. Did you compose anything during these past few years?

YALMAR. Yes... You'll hear. But you'll be the only one. We'll choose a time when all the others have gone for a walk.

PAULA. But Count Schperling is a music lover.

YALMAR. That's why I'm not going to play for him. Those who don't understand anything you can endure, but those who imagine they do are unendurable.

SCHPERLING. Miss Paula, maybe you, will grant us the pleasure of hearing you play before the rest of the guests arrive. Yalmar and I are both very fond of music.

PAULA (*hesitating*). I don't think the baron wants to …

YALMAR. Listen…? On the contrary, I'm very glad to.

> (*He offers his hand. Schperling follows after her.*
> *Selen hurries ahead and opens the door.*)

ALISA (*to Aunt Amelia*). And to the old women, no one offers a hand.

AUNT AMELIA (*going along with her*). It's really quite insensitive on the part of this girl.

ALISA. They don't even take us into account – either of us – Aunt, dear. (*She inclines her head towards her shoulder.*) Ah, I feel so old, so superfluous.

> (*They enter the castle just as music is heard*
> *coming through the open windows.*)

ACT TWO

An open area in front of the factory. On the right are factory buildings. In the depths of the scene, on the left, is a waterfall, almost dry, with a large wheel that is not operative. Another wheel connected to it stands closer to the factory buildings. There are stacks of supplies and such. The castle can be seen in the distance on the right. On a platform on the right are benches and chairs for gentlemen. On the left there is a podium for speakers made from stones stacked on top of one another. When the curtain rises, workers and factory employees are beginning to gather.

ANDERSON (*walks in with some workers, talking heatedly and waving a paper in his hand*). Look! I've figured everything out to the penny. I can tell you how much he stole from each of you. Let's take you, Sven Karlson. With a family like yours, you would have saved 50 krona or more this year if we had charged the same prices as the shop in Lido. This is what you should be getting today in the form of a dividend. Let's see if you get even half of that!

SVEN KARLSON. You're right! Why should I lose money?

ANDERSON. Just look! You don't have to agree to less than 50 – I'm telling you.

ERIK. What did he do with the money?

ANDERSON. I don't want to say anything – but have you ever been to his laboratory?

ERIK. Yes. It's full of expensive things.

(*Alisa comes out of the castle with all her guests from the previous act. They're engaged in lively conversation as they sit down on the benches.*)

ALISA (*to Count Schperling*). I totally share his opinion. The workers' only hope of improving their conditions is to stand together.

SCHPERLING. But it's a shame that all these *friends of the people* have such peculiar personalities?

ALISA. That's what people have always said about people who have new ideas.

BARON KRONSTREM. And they speak the truth, quite accurately, Baroness. Intelligent, conscientious people never engage in the spread of new and immature ideas.

ALISA. So, in your opinion, the world should always stand still – never move forward?

KRONSTREM. Of course not. Society will move forward without any directives. These young innovators only make matters worse when they talk about wanting to move the world forward. The only thing they're really interested in is their own advancement.

(Stenson appears among the workers.
Martha approaches Alisa, who greets her.)

ALISA. So, you're also interested in consumer partnerships?

MARTHA. No. But I jumped at the chance to come with my father. I love being here. And I wanted to see Paula. Are you going to introduce me to your guests?

ALISA *(gesturing towards Martha)*. Miss Stenson.

MARTHA. Alisa and I have been friends since school. We love each other so much that we don't pay any attention to the fact that our men are in the middle of a dispute.

MDM SELEN. This means it was your father...

MARTHA. ...who beat the baron to the development of the dams? Yes. But we women don't have anything to do with that business – do we? You've heard there's a big scandal going on today! My father says the workers are extremely angry with the engineer.

SCHPERLING *(laughing)*. They are? Did you say extremely angry?

ELFORS. It does seem rather strange – that the workers would feel distrust towards a person who's shown so much concern for their welfare.

ALISA. He just can't identify with them. That's all.

(Paula comes out on the other side, talking briskly with Yalmar.)

MDM ELFORS *(to Countess Schperling)*. This girl must be a terrible charmer, even if on the surface she seems to be the embodiment of innocence. How did she manage to turn the head of the baron in such a short time?

COUNTESS. Yes, can you imagine? Yesterday – when we all went skating – and Baron Yalmar said he couldn't go with us on account of the workers – business – remember?

MDM ELFORS (*curious*). Yes, so what?

COUNTESS. Well! As soon as we left – my maid told me – Miss Torell came to the castle. The baron met her on the stairs and they went straight to the concert hall, where they stayed all evening.

MDM ELFORS. I see. But they never played when we were there.

COUNTESS. And my husband is so musical!

YALMAR (*to Paula, quietly*). You go on! In a minute, I'll follow. Everyone's so interested in the scandal that's about to unfold that they won't even notice us.

PAULA. But I'm afraid Alisa will get the wrong idea. Let me tell her.

YALMAR. Please! What are you thinking? Do we really have to put every word, every look into a record and submit it to Alisa for review – even if this *is* the generally accepted idea of how married people ought to behave?

MARTHA (*runs towards Paula and hugs her*). *You*, naughty girl, have been home for a whole week and haven't even come to see me!

PAULA. I'm sorry, Martha, dear! But with the way things are between your father and Karl... Your father literally torments Karl.

MARTHA. Karl is the one to blame. Why is he so stubborn? My father can't work with unions and things like that. There will never be anything like that at Lido – no partnerships, no religious sects, nothing of the kind.

PAULA. But what business is it of your father's, if here at Gerrgamra, Karl...

MARTHA. But I have to deal with this. Why is Karl so selfish? Why does he think only of himself? As long as my father is angry with him, Ernest and I will never be allowed to get married. If only Karl would agree to my father's offer of a manager's position at Lido – as we all know he should!

PAULA. Karl has more important things to do.

(Ernest enters. Martha runs to meet him.
They go to one of the factory buildings to kiss.)

PAULA *(to Alisa)*. I hope you won't mind if I disappear for a minute with your husband. We were hoping to play a little, while nobody's in the house.

ALISA. And you ask my permission? As if you care about what I think.

PAULA. Alisa! If it's upsetting to you, we won't go.

ALISA. Please. I really don't care. But how can you think this little tryst you're planning is so important that...

PAULA. No, no! Not another word. You're right. I'm staying.

ALISA. But not for me, remember. Also, he's still waiting for news from the academy about the prize, isn't he?

PAULA. Yes, a friend wrote to him about it. He wanted the meeting to open an hour later, so that he could get the mail. This would give him more courage and strengthen his authority.

ALISA. That's true. *(She goes and makes a sign to Ernest to approach.)* Would you like to do your brother a great service?

ERNEST. Of course.

ALISA. Take a horse and go to meet the mail. It would be helpful if your brother got the news from the academy before closing the meeting.

ERNEST. Yes, of course, I will. I'll ride straight through the fields. I'll be back in half an hour. *(aside, to Martha)* You know, babe, I have to leave you for a minute. Be smart, now, and don't fret over it. *(He kisses her stealthily and hurries away.)*

PAULA *(moving aside with Yalmar)*. Alisa doesn't like it.

YALMAR. Doesn't like...? Oh, well...! Without her permission, it's out of the question.

KARL *(enters accompanied by Mdm Torell. All look expectantly at them. He turns to Paula.)* Where's Ernest?

ALISA. I sent him on an errand. He'll be back in half an hour. Do you need him?

KARL. Yes, but that doesn't help anything.

MDM TORELL. Karl wanted to ask Ernest to go to the post office. He is expecting to hear about the grant.

PAULA. That's why Alisa...

ALISA (*carries her aside*). Don't say anything.

PAULA. Why?

ALISA. Well, just tell him that Ernest decided himself. Don't mention me.

MDM TORELL. There's been such agonizing anticipation all these days. But, now, at least we know that the academy meeting was yesterday. His whole destiny depends on this. Imagine! To do so much work and not be able to continue! That would be terrible.

KARL. Let's not bother the baroness with this. It's all well known to her.

MDM TORELL. Yes, but I can't help but tell about how painful it is to me that my poor boy, who's already experiencing such anxiety, is being further troubled and tormented by all sorts of other things.

(*Karl wants to interrupt her.*)

MDM TORELL (*shakes her head*). He's become so nervous. It's almost unbearable.

STENSON (*coming to him*). Well, my young friend, is it true that you've finally finished your work?

KARL. On the invention...? Yes.

STENSON. When will we see this new machine in action?

KARL. When I get the funds needed to build it.

STENSON. Ah! More money! Well, then you'll have to look for some other way to turn a profit than this consumer partnership.

KARL. As to whether it's been profitable, you'll see. Are you ready? Can I begin? (*rises to applause*) I have called you here today to share the result of a report on our status, which has been approved by the Audit Commission. Unfortunately, the result was far from what we all expected. (*There is murmuring in the crowd.*) Not everyone joined the consumer partnership, not everyone trusted the plan – and this was the only way we could ensure the success of the new enterprise. You – most of you – preferred to buy bad goods at the Lido shop.

(murmuring)

STENSON. Let the buyers judge the quality of the goods themselves.

KARL. You've allowed yourselves to be caught up by deceptively cheap prices.

VOICE. Deceptively!

KARL. Deceptively, I say, because Stenson sold at a loss to attract buyers.

VOICE FROM THE CROWD. If that's the case, it was right kind of him!

(laughter)

ANOTHER VOICE. Yeah, we have nothing against that.

KARL. But you have to understand that prices at Lido will not stay that way. It's just a ploy to strangle our consumer partnership. That's the whole point.

ANDERSON. We would ask the chairman to stick to the case and not sidetrack.

KARL. I'm not sidetracking – because, in essence, Stenson's entire effort has been to convince the members of the partnership that, should the first year be unsuccessful and not up to expectations, then you should look for reasons not to trust the whole idea of cooperatives and not to trust me.

ERIK. Why would Stenson want to destroy our partnership? We don't understand.

KARL. Because Stenson is an ardent adversary of all workers' unions. He fears the workers will become a force.

STENSON. And Torell is the defender of all kinds of workers' unions, because he hopes to create a political force with the help of them and then to climb on their shoulders. Just wait for future elections in the Riksdag and you'll see.

(laughter)

MULTIPLE VOICES. Wrong assumption!

KARL (*strikes with a hammer*). I call the assembly to order.

VOICE. Well, give us a quick report. How much will we get as a dividend?

KARL. As can be seen from the reports reviewed by the Auditor, no dividends will be issued this year. But...

A LOT OF VOICES
> He blew us off! Our miserable labor was for pennies!
>
> He stole from us!
>
> We know where our money's gone!
>
> He's got a whole lab full of models!
>
> And they're worth a whole lot of money!

ALISA (*quietly to Yalmar*). You have to say something and be firm in his defense.

YALMAR. I? What do I care about this?

ALISA. You know how unfair this is! You have to tell them where he got the money! And if *you* don't want to speak up, I'll do it!

YALMAR. You're so eager to be ridiculed?

ALISA. I don't care! Only a coward could hear such slander and keep silent!

KARL (*when the noise has somewhat quieted down, strikes with the hammer to call for silence*). In light of the shameful accusations being made against me, there is only one thing left for me to do – to call for order and ask that the assembly choose another chairman.

ALISA (*trembling under the tension, takes a few steps forward and speaks in a shaking voice*). Mister Chairman... (*Everyone is astonished. The people on the benches rise.*) May I say a few words?

KARL (*looks at her. Their eyes meet.*) A word from Mdm Baroness Yullenyelem.

ALISA (*holding on to the stand and speaking somewhat nervously*). I apologize if I fail to choose the perfect words to express my thoughts... I have never before had a reason to speak publicly. But, since no one wishes to speak in defense of principle – *(beginning to speak in a more confident tone)* – in defense of one who is being subjected to unfair accusations – I find that it would be an unforgivable weakness on my part to be silent. *(Looking at Paula, Alisa begins to look calmer.)* You all – fathers and grandfathers, who have worked here at the Gerrgamra factory – you all know me. You know that I would never tolerate unfairness towards you. You know that I would never do anything to protect a person who would dishonestly dispose of the money you have earned – who would do what you accuse him of. You wonder where he's getting the money for the experiments, so I feel obligated to tell you. He was given this money as a loan from us – from my husband and me – because we believe in his invention – from which, someday, you will all benefit. *(She stops for a minute, takes a breath and drinks from a glass of water brought to her by Schperling.)* Yes, I feel that I must say this. You may review the reports yourselves, or refer them to people you trust, but you should not throw such shameful accusations in the face of a person who is working to benefit you.

(She quickly moves away and hides among the guests.)

PAULA *(runs after her and hugs her)*. Alisa! You're a heroine!

ALISA *(removes herself, agitated by her voice)*. If you hadn't attracted so much attention from Yalmar, he might have saved me from this.

STENSON *(to Anderson)*. Well, now that the meeting has taken such a – shall I say *romantic* turn – *(makes a bewildered gesture with his hand)*. I find it impossible to embark on such a prosaic action as the election of a new chairman, or a reading of the report, or anything else. *(whispers)* Move to close the meeting.

ANDERSON *(shouts)*. I propose to close the meeting for today.

(fervent speaking, passing from one group to another)

KARL. Do we accept the proposal by Mister Anderson to close the meeting?

UNANIMOUS WHOOPS. Yes!!! Yes!!! Yes!!!

KARL *(strikes the hammer)*. I declare the meeting closed. *(quickly goes up to Alisa)* How can I thank you? You did this for me without getting any concrete proof that I'm not a crook – not a charlatan.

ALISA. I did it out of a sense of justice. You have nothing to thank me for.

MARTHA (*shouts to Alisa*). Ernest is coming! Do you see him? He's riding so fast! He's completely driving the horse!

KARL (*looks in the same direction*). Ah, so that's why...

ALISA (*embarrassed*). Your brother wanted it.

(*Karl goes to meet Ernest, who is approaching, waving a mailbag.*)
(*Baron Yalmar opens the bag Ernest hands him and hands over to Karl a large envelope. Karl takes it, quickly. All eyes turn towards him.*)

MDM SELEN (*to Miss Kronstrem*). This must be the news about the grant award.

MISS KRONSTREM. The news our hostess is so interested in!

MDM SELEN. She was so adorable when she stood up there. I never thought a lady speaking could create such a charming spectacle.

MDM TORELL, MARTHA, ERNEST and PAULA (*to Karl, who lowers his hand with the letter, convulsively crumpling it*). Well...? Well...? Well...?

KARL. You threw your 15,000 into the bottomless pit, Baron Yullenyelem.

(*There is a pause.*)

YALMAR (*rising*). Well, then, I'm very sorry – for your sake. As for me, to tell the truth, I have long since considered this money lost, so please don't worry about it. I only hope that you're cured now, at least.

KARL. I understand that what I am about to say may seem ridiculous at the moment, but I will still have the satisfaction of saying it – although I know perfectly well that there's not a person in the world who will believe me now. Baron Yullenyelem, you are wrong to consider your money lost, because the problem does have a solution. I probably won't live to see it – but it doesn't matter. I may have to accept the thought that I'm going to live and die regarded as an adventurer or a fool – but my invention will long outlive me.

(*He starts to leave.*)

ALISA (*runs after him*). Karl!

(*He stops. They stand apart from the others.*)

ALISA. I believe you.

KARL (*grabs her hand*). Alisa! What does this mean? (*quietly*) Is this also spoken out of a sense of fairness?

ALISA. Karl, I love you. I will follow you wherever you go. I will give you everything I have. And, together, we will succeed.

KARL (*shocked, kisses her hand*). It's too much – too much happiness at such a terrible moment. I'm not able to understand.

(*Alisa leans and kisses him.*)

YALMAR (*runs to her, beside himself. The guests follow him amazed.*) Are you crazy?

ALISA (*takes Karl's arm and looks into Yalmar's face*). Say what you want, but we two are together now.

ACT THREE

About six months after the previous act. Karl's lab. Evening. In the middle of the room, on the table, is a lighted lamp near the electric device. Models of different kinds of machines are arranged around the room, which is furnished with just a simple table and some chairs. In the background is the door to the factory office. Outside, the ground is being covered with snow mixed with rain and there is a strong wind.

ERNEST (*enters with Martha through the office door. Both are covered with snow and wet. Martha is wearing a scarf on her head, a coat, a fur cap and boots. Ernest has a raised collar and a lantern in his hand.*) Hurry, I'll show it to you before Karl comes back. He's been so afraid of spoiling anything that hardly anyone has seen the device. He's already thinking...

MARTHA (*rearranges her scarf and goes to the machine*). So, this little gadget can make the whole machine move.

ERNEST. Yes, that's it! That little thing! How many days we've thought about it! How many hours we've worked on it and struggled. You can't imagine! Often, for nights in a row, Karl didn't even bother to undress. But all we have left to do now is to make a tiny adjustment and tomorrow morning the machine can be put into action.

MARTHA. And then Papa will want to manufacture with Karl and we...

ERNEST. We will get to celebrate our engagement! What did your father say about the idea of working here at the waterfall? Did you hear anything?

MARTHA. Oh, he was talking to the workers and he said he'd have fun when fifty people got fired.

ERNEST. Did he say that? Well, now it's clear why they all started worrying. And Karl hasn't come in! I think I'd better go back down and have a look. What if there's going to be a riot?

MARTHA. I'll stay here, and, if they come, I'll throw stones at the window!

ERNEST. I think you'd better come with me.

MARTHA. But what if they start fighting down there? No, I don't dare. I really can't decide. What should I do? Oh, why, did I even come with Papa today?

ERNEST. To see me, my sweet fiancé, I hope. But you haven't even really looked at all the models, yet. Look. They're almost like beautiful, meticulously made toys. We're going to give these kinds of toys to our children someday, right?

MARTHA (*smacking him on the cheek*). Oh! You're being silly! Yes, they're very pretty! But didn't all this cost a lot of money?

ERNEST. Oh, most definitely.

MARTHA. But where did Karl get the money? Alisa had nothing when she left Gerrgamra.

ERNEST. No, she had a small cash inheritance from her mother. But it's long since dried up. We had to cover so many failures and accidents. Now, we live on the money we get from the sale of her diamonds, her old silver, etc. Take this machine, for example. Do you know what it really is? It's Alisa's beautiful diamond necklace – a part of her mother's legacy. And this here – guess what it is. You may think it's a miniature locomotive – but, no! It's, actually, the large silver vase on high legs that was always the centerpiece on the table when Gerrgamra held great parties to celebrate the opening of the hunting season.

MARTHA. That's very touching, you know – truly, very touching. I wish I could donate my jewelry, too. Yes, I would love to. I have a gold brooch with a diamond. Do you think it's possible to turn it into a locomotive – just a tiny little locomotive?

ERNEST. Oh, no, you don't need to part with your jewels, my love. We're finished! It won't be long until Alisa will get back everything she's sacrificed.

MARTHA. I noticed that you were denying yourself, too, sweet boy. You always ate so hungrily whenever I asked Papa to invite you to lunch on Sundays at Lido.

ERNEST. Yes, I remember. Such juicy, delicious roast beef. Ah! How my mouth watered!

MARTHA. And, when you were embarrassed to take seconds, I put a big piece on your plate and your eyes lit up. You'd think I had offered my lips, instead of a piece of roast beef!

ERNEST. Oh, I don't know if even they would have been as tasty.

MARTHA. Shame on you! Just for that, I won't let you kiss me for a whole... *(He interrupts her with a kiss.)*

ERNEST. But, now, we're penniless. If we should happen to have any sort of problem with the machine, we don't have enough money for even the smallest repair.

MARTHA. But Papa always wonders why you don't live on the income from Gerrgamra.

ERNEST. We can't touch it, because we know we'll have to return everything if the settlement decision doesn't come down in our favor. Karl has had to put all income from the factory in a trust. It's being held in escrow for the future owner.

MARTHA. But, who will the owner be? Do you really think Baron Yalmar could win the case?

ERNEST. Oh, no, he probably won't win. Most likely Gerrgamra will be recognized as an in common hereditary estate and be put up for sale... and then Karl will buy it. Do you see? When he sells the patent for his invention to countries all over the world, we'll have as much money as the sands in the sea! Shh! Karl's coming up the stairs. For heaven's sake, don't touch anything.

KARL *(accompanied by Alisa, both in their coats, turns in alarm to Martha, who is standing near the apparatus)*. Oh, my God! I hope you didn't touch anything!

(He runs to the apparatus and looks. Everything is in place.)

ALISA *(to Ernest)*. I'm so worried. The workers aren't going home today. They're crowding around the waterfall, whispering and looking at the battery device. And Stenson and Anderson are weaving in and out among them, stirring up trouble.

KARL *(engaged in the apparatus)*. If only we could have the demonstration ready for tomorrow morning. Then they will stop. Even the most inexperienced workers will be amazed when they actually see it in action.

ERNEST. Look, Karl! Mother sent me here to ask you and Alisa to come home for dinner. She's preparing a special supper to celebrate the battery in a grand way. You'll never guess what delicious things she's managed to get! And Papa Stenson is even allowing Martha to spend the night with us, because of Paula's arrival.

KARL (*leaning over the machine*). It's too early to celebrate. Wait until the machine is up and running.

ERNEST. But once it's been installed and everything...

KARL. Besides, I don't have time to eat today. I'm not leaving here until everything's ready. I'll have to work at least all night.

ERNEST (*feigning helpfulness*). Maybe I should stay and help you.

KARL (*smiling*). And miss out on the gala dinner. Oh, no! That would be too cruel! Besides, I don't need you. I don't need anyone today. I only need one thing – to be left alone.

ERNEST (*taking Martha's waist*). Come on, sweet girl. And you, Alisa?

ALISA. I'll stay, of course.

ERNEST. But didn't you hear what he said: I don't need anyone. So, you're no one to him even if you stay. That's great! Then, Martha, maybe you're no one to me, too.

MARTHA. No one! Aren't you ashamed? (*They go out together.*)

ALISA (*leaning over Karl*). Ernest was right, you know. I'm no one to you. I can't be another person to you – because you and I are one person.

KARL (*holds her*). It feels strange to have finally reached our goal.

ALISA. Yes, it does feel strange. I even feel a kind of emptiness at the thought.

KARL. Oh, I'm full of plans for future work – enough for a lifetime!

ALISA. But none of these future works can do for us what this did – our beloved brainchild – the dream that brought us together and for which we had to suffer so much. Now, you're going to earn so much money that we'll have no economic difficulties. And – can you imagine – the prospect is almost unpleasant to me. I would rather that we could always live in our modest little home. I don't think there's a happier home in the whole world. *(They hug one another.)*

KARL. Can't I even be glad that you'll no longer need to deny yourself anything – that once again you'll be able to enjoy the comforts of the glorious lifestyle you had grown used to from childhood?

ALISA. Don't give that any more thought, my love. I assure you that doing without was never stressful to me. In fact, it seems to me that a person has more right to enjoy personal happiness when he's denied himself something in exchange for it. It's impossible to take all this in – to think that, in addition to all the happiness we've had, we will now have great riches. It pains me to think about it. It will only increase the gap between the workers and us. The reason I've been so happy, all this time, is that we lived no better than all those around us. The poorest of our workers had no reason to envy us. And when I think of our little one, who will soon be born, I find myself wishing he could grow up among the other children as an equal, not as an exceptional person, like me. I hate thinking of him being born a wealthy man's son.

KARL. It won't hurt him, if we nurture him so that someday he'll build the palace for the people you've been dreaming of.

ALISA. You mean – equip! Because, it's already built. The big hall will be turned into a school – the billiard room, into a lecture hall – the library, into a reading room. I've been thinking so much about these plans.

KARL. Well, that's fine! Just let it all go!

ALISA. That's precisely why I want our son to be born under our present conditions, or else, it will be the same with him as with you. Your mind and your innate sense of justice draw you to one side, but your instincts to the other.

KARL. More correctly – my experience. When you become as familiar with the people as I now am, and you see their incorrigible, idiotic resistance to all that you want to do for their benefit. No nobleman is so conservative, so suspiciously opposed to all innovations, as these people. No, I'm quite cured of all my former democratic tendencies.

ALISA. But, then, what joy is there in having invented something, if it won't contribute to people's happiness?

KARL. It will, of course, contribute, but in its own way. Gerrgamra will become an important industrial center over time. There are many pessimists who think our country can't long survive, because our agricultural system is failing. I'll prove to them that, if we're able to harness the electrical power our waters can produce, we'll be able to manage the use of our other resources, and, in the course of time, we'll be in a position to compete with the largest industrial nations.

ALISA. But what's the benefit of that? Poverty is as much a problem in large industrial nations as it is in ours.

(There's a knock on the door.)

KARL. Come in!

SENIOR FOREMAN *(enters)*. Excuse me, but may I ask you a question?

KARL. Not if the question can wait. I'm too busy right now.

SENIOR FOREMAN. I fear this question cannot be postponed. People are worried tonight. And, since I'm the senior foreman, I will have to answer for it if they begin to riot.

KARL. Of course! You're too weak! I've said so many times! You exercise no discipline!

SENIOR FOREMAN. Excuse me, Mister Engineer. But, when people are starving, bread is more important than discipline. They had to endure a lot because of the lower wages.

KARL. No one here can help, until it's been decided who will be the factory's real owner. The settlement process will be over soon.

SENIOR FOREMAN. But, in any case, the workers say that the engineer is guilty of…

KARL. Of what? Say it!

SENIOR FOREMAN. I'm uncomfortable talking about it in front of your wife, but the fact is the workers can't understand all this. They say that, if Mister Engineer – if I may say so – had not distracted the baroness from her duties, this misfortune would never have happened. If the baron were still the owner, the factory would still be working.

KARL. What does that have to do with what you're talking about now?

SENIOR FOREMAN. Well, you see, of course, that this is all related to their hatred for the battery. They think that you, Mister Engineer, invent things that will lead to their dismissal. And that's why I came – to ask you if what they say is true, or not. Can I, on your behalf, go and reassure the workers and tell them that all these stories that make claims, such as saying that you intend to dismiss fifty workers this spring, are lies?

ALISA. How can they think...?

KARL. Those who behave well and perform their duties will remain at the factory. But those who prefer to make noise and revolt will be fired. You can tell them that!

SENIOR FOREMAN. So, I can't tell them that there's nothing to worry about and that they should all go home?

KARL. No! No promises. I will introduce new methods of work and those who can't learn to apply them will no longer be needed. In any case... Yes, it will be necessary to reduce the number of workers.

SENIOR FOREMAN. So, the people are right, when they say that the new machine is a threat to them.

KARL. Yes, for those who are in the habit of resisting change, it will do nothing for them. But, in general, it will produce a useful shake up among the workers. Can you talk to the workers and explain this to them? Most importantly, you need to convince them to go home, for now.

SENIOR FOREMAN. All right... I'll try. (*He leaves.*)

KARL. I don't seem to have a chance at being alone for even one minute. And I really must get this machine ready to demonstrate by tomorrow morning.

ALISA. You never told me about this, Karl! Dismissing so many workers!

KARL. So, that's it! Now you'll stop supporting the invention.

ALISA. But, fifty workers! Just think how many of them have had permanent residence here – ancestors and fathers of whom lived in the same house as they during the lives of my father and grandfather.

KARL (*with irritation*). Is it absolutely necessary to talk about this now, when everything depends on whether I can work quietly? (*adopting a more friendly tone and reaching for Alisa's hand*) I'm sorry, my love – only – please, not now.

(Martha rushes in, followed by Mdm Torell and Paula. Karl throws them an impatient look.)

MARTHA. Alisa! Guess who's here! Baron Yalmar! He's just arrived! We saw him climbing out of his sled. I assure you, it's true. We saw him – though he didn't notice us, because Paula ran away and dragged me along.

ALISA (*in an agitated voice*). Oh?

(Karl pushes his chair back.)

MARTHA. Think how exciting this is! Promise me that I'll be here to see it when you two meet again for the first time!

(Mdm Torell tries to silence her. Karl looks at Alisa with searching eyes.)

ALISA. It's nothing. Yalmar and I will always be able to meet like old friends.

MARTHA. But isn't he mad at you?

(Karl reaches for Alisa's hand. Alisa approaches him, puts her arms around his neck and leans her head towards him.)

KARL (*quietly*). Your cheeks are burning. Does this bother you?

ALISA. It bothers me that he's shown up just now, when I'd like to be thinking of nothing other than your big moment.

KARL (*lets go of her*). This is what happens when we have a past that has such a strong influence on us. (*He gets up.*)

ALISA (*wraps her arms around his neck again*). I don't want any past to influence me.

MDM TORELL (*sets the table with Paula's help and prepares a supper*). Come, let's eat.

KARL. What's this? I don't have time.

MDM TORELL. But you can't go hungry all night. I won't disturb you for long, but, if you're thinking of staying here all night, you should definitely eat something.

KARL (*eats a sandwich standing up. The others sit down.*) Where's Ernest?

MDM TORELL. He went to try to appease the workers. They became terribly excited at the sight of Baron Yalmar, immediately spreading the rumor that he had won the settlement. They started screaming and cheering! It's disgusting to hear them! Such ingratitude! I hurried past them as quickly as I could. It's an unpleasant business.

(She says something quietly to Karl.)

ALISA (*takes Paula aside*). Do you think he came for you?

PAULA. Yes. I think so. As he was leaving Stockholm, I sent him a note where I wrote that I was coming here to say goodbye to all of you, because I intend to take the offer of a concert tour in America.

ALISA. So, you just saw him, in Stockholm?

PAULA. Yes, I did. He was at my concert and he brought me a magnificent bouquet of roses – with drops of dew made from diamonds!

ALISA. I don't understand him. He loves you. I'm sure of it. And, yet, he wants to be the owner of Gerrgamra, even though he knows that, as the owner, he would not be allowed to marry any person whose ancestors were not noblemen.

PAULA. I understand that very well. He's too aristocratic, you see. Aristocratic instincts are in his flesh and blood, and he's, also, too accustomed to not working. He can't envision how he could earn a living. You can't imagine what a struggle he's been going through all this time. I could see it, though he tried to hide it from me. But I kept hoping all this time that he might, for example, take the place of the conductor of the orchestra in the great opera in Stockholm.

ALISA. Do you think he could be content with such an occupation?

PAULA. Yes, in fact, I think he would be a lot happier and better off. The wildest hopes have been stirred up in me! The fact that he left immediately after me... *(hugging Alisa)* Oh, Alisa, if only it could happen!

(There's a knock on the door. Paula and Alisa bounce off each other with a suppressed exclamation. Karl jumps from his chair.)

MARTHA *(screams)*. Oh! God! It's him!

KARL *(trying to give confidence to his voice)*. Come in!

(Erik enters, slowly.)

KARL. It's you? What does this mean? You've left your guard post?

ERIK. Yes. I came here to tell you, Mister Engineer, that I don't want to be on guard tonight. I think, as do the comrades, that all this battery is going to accomplish is the demise of the workers, and I don't want to have anything to do with it.

KARL. I had ordered you to stand guard for two more hours. At that time, you'll be replaced. I don't intend to talk to you about what you think, or what you don't think. Go and do your duty, immediately – that is, if you don't want to be fired tomorrow.

ERIK. Then, I will ask the lady, who has always wished only good for the people...

KARL. No mistress gives orders here! I do! I say, go!

ERIK. If I stay at guard, the comrades will give me trouble.

KARL. Oh, so that's it! Well, we can defend ourselves against violence! Let them just try to rebel! I'm ready to call the police, now! In the meantime, you'll fill your position until you are relieved. Or else I, myself, will go stand watch, and you'll be dismissed tomorrow morning and thrown out of your apartment.

ERIK. Thrown out...! From the apartment where I was born, where my father lived before me...?

KARL. There's no excuse for acting disrespectfully and neglecting your duties.

ERIK. And the girl, Marya, who can't stir because of her sore leg... and my wife, a childhood friend of Alisa's as a young lady. Can't the baroness put in a good word for me?

KARL (*interrupts Alisa, who wants to talk*). The baroness does not exist, anymore, as you know… and Mdm Torell will never have a good word for someone who has shown such disobedience to her husband. Of this, you can be sure! Well, are you going to perform your duty, or not?

ERIK. The baron has returned home, and, now, it seems, the engineer will not have a long time left to be in command here.

KARL. Well, then… Get out of here!! (*Karl puts on his coat and turns to Alisa.*) Please be sure that everything is left as it is and properly locked and that the light is extinguished. It seems I'm going to be keeping watch myself tonight.

ALISA. Erik, think about Anna and your poor, sick child… and don't tie your hopes to the baron. The engineer is in charge of the factory, now, and the baron has no right to interfere.

ERIK. But they say the baron has won the settlement process.

ALISA. The court hasn't handed down any decision.

ERIK. No decision? But the comrades think… What if he should come in here right now?

ALISA. Tell them, it's a mistake.

ERIK. And what should I do if there's still trouble?

KARL. If you wish to do your duty, then send me an immediate signal by the electric light. You know how to operate it. I'll be here all night and I'll come as soon as I see the light.

ERIK. Well, all right… I'll go!

KARL. Remember, you will answer to me if anything happens. Here's the key to the shed over the battery. There's a screw inside. Just turn it and the light will immediately turn on. And, understand that, if you should *not* call me as soon as there's any danger to the battery, nothing – no baron in the world, you hear! – will be able to save you from the fact that you will be fired and thrown out of your apartment in the morning! (*Erik leaves.*)

ALISA. You shouldn't have spoken to him in such a sharp, imperative tone. It's really no wonder they find it infuriating.

KARL. And perhaps you'd like them to destroy the machine – just at the moment I've finished it – after all this hard work.

ALISA. They would never have acted so menacingly, if you hadn't pushed them so harshly.

KARL. Oh! Well... What can I say? I have every reason to be friendly to them after the way they've treated me all this time. You didn't protect them last year when they showered me with insults. On the contrary, you followed the prompting of your heart and came to me and said: I believe you. But, now, apparently, you don't believe quite so much.

ALISA. Oh, Karl! I know you want to be good to them. Only, you have such a sharp manner.

KARL. No! I don't want to be good to them. I just want to be left alone.

ERNEST (*rushes in*). I think you should come, Karl. They're starting to spread out. The arrival of Baron Yalmar has encouraged them. They think he's going to support them if they do some harm to you or the machine.

KARL. So, you think they're ready to carry out some kind of violence?

ERNEST. Yes. I'm afraid they are.

MDM TORELL. Oh! Lord, help us! What are you saying?

MARTHA (*clinging to Ernest's hands*). What do we do?

KARL (*to Ernest*). Let's go downstairs!

ALISA. Are you going, Karl?

ERNEST. Oh, Alisa, don't be worried.

ALISA (*to Ernest*). I think you'd better go, Ernest – in case Karl's presence just makes things worse. You just have to talk to them in a friendlier manner.

ERNEST. Be friendly? With this scum – with these ungrateful scoundrels?

KARL. Alisa, here, has nothing but friendly feelings towards them.

ALISA. I only want one thing – to avoid any disturbance. You go, please, Ernest, and try to calm them down. It will be so awful, if it comes to violence. Ask them, on my behalf. They're loyal to me.

MARTHA. But Ernest mustn't leave *me*!

ERNEST. Yes, maybe it's best if I take the women home first.

MDM TORELL. And we should then just sit upstairs, alone! What if they come and attack our house! Isn't it better if we stay here?

KARL (*with irritation*). Then say goodbye to all hope of getting work done tonight.

MDM TORELL. Yes, you really should not be disturbed.

ERNEST. You'd better come with me. I'll ask one of the office workers to sit with you.

MARTHA. But I'm afraid to leave. Can't I stay here? I'll be as quiet as a mouse.

KARL. Stay, or leave, as you please. Just do me one favor and go out to the office and let me, finally, get back to my work.

(*He sits down to begin work.*)

MARTHA (*whispers to Ernest*). I'll stay in the office. Just come back soon.

(*Mdm Torell and Paula go with Ernest, through the office door. Alisa lights the lamp and sees Martha into the outer office, leaving her there. She, then, returns to Karl, who doesn't raise his eyes, even though she's standing near him.*)

ALISA. Maybe I can help with something?

KARL. No, thanks! Actually, I think you should go home. You haven't slept much all these nights and you shouldn't be overtired now.

ALISA. So, I'm also bothering you?

KARL. No, it's not that… but (*looking at his work*) maybe you want to see Yalmar?

ALISA. What does that mean, Karl? You're insulting me!

(*She leans over him.*)

KARL (*removes himself*). Yes, I know I insulted you. There's nothing I can do about it. I'm so irritable now. I'm in such a nervous state. I'm so tired. But understand that it's very hard to stand at the goal, after such serious, long work and realize that, in the decisive moment, the one who means the most has withdrawn from you.

ALISA. You have no, right, Karl. How can you say such things?

KARL (*his voice is breaking with emotion*). You have lost all interest in my work. You only think about the workers' grievances and you stand on their side against me.

ALISA (*hugging his neck with both arms*). Oh! My love! What I wouldn't give to be entirely on your side! But you're so focused on your scientific interests that you've forgotten the real goal.

KARL. I have only one goal at the moment – to make this machine work, as it should. I can't think of anything else right now. (*He puts his hand to his forehead.*) Just look! My sweat is pouring out in buckets. There's a knocking and a kind of bubbling in my brain! The machine *must* be put into motion tomorrow – even if it means the death of me!

ALISA (*runs to Martha*). Come on, Martha. We're going home. Karl really needs to be left alone.

MARTHA. You and I, leave here, alone? Are you crazy?

ALISA (*losing her temper*). Be quiet! No noise! Hurry up!

> (*She throws a coat on herself, helps Martha put on her coat, takes in her hand both pairs of boots and pushes Martha through the door ahead of herself.*)

ACT FOUR

The same scene as in the second act, only it's evening. It's dark. The noise of the waterfall is audible. A large wheel, previously driven by the waterfall, is connected to a battery located in a shed. The scene is illuminated only by fire from the blast furnace and light from the window of Karl's laboratory, located on the right. There is wet snow and wind. Workers are gathered near the waterfall. They are not clearly visible, only audible, as they quietly whisper about something.

(Alisa and Martha enter from the right, huddled together. They are moving slowly, often sliding, almost falling. Their scarves cover their faces. Alisa illuminates the road in front with a lantern. Suddenly, they shudder and cry out, as they stop and peer into the darkness.)

MARTHA. Psst!

ALISA *(whispering)*. What is it?

MARTHA *(also whispering)*. Don't you hear voices?

(They stop and listen.)

ALISA *(as before)*. I think it's just the noise of the waterfall. You're just not used to it, yet.

MARTHA. No, I heard voices – rude, frightening voices!

ALISA. Do you think it could be…? Maybe we should go back.

MARTHA. Yes, let's go, now! Karl should escort us! He needs to stop thinking only of himself. He was supposed to take care of us. Listen! Come on! *(She pulls Alisa's hand and wants to turn back.)*

ALISA. Shh! *(She looks round in fright and whispers.)* Did you hear footsteps behind us?

MARTHA *(muffles a scream and grabs for Alisa)*. Put the lantern out so we don't stand out.

ALISA. Then we won't be able to find our way home.

(She raises the lantern. The light falls on Anna, who approaches them.)

ALISA. Who's there?

ANNA (*nearer*). For heaven's sake! I hadn't seen you. You frightened me, Baroness! Why would you be in the factory yard in this weather?

ALISA (*at almost the same time*). Anna?! What are you doing here tonight in such darkness?!

ANNA. I came running to look for my husband. Surely, you know, Baroness – or I meant to say, Ma'am – that the engineer has ordered Erik to stand watch over the batt…beasterie… or whatever it's called… this ill-fated new machine. And, now, things are so bad that some of the worst workers have threatened to do something to him if he doesn't abandon his guard. I was troubled by such fear at the news that I rushed out to get here, although it was difficult to leave our girl, who's doing very badly. Let me help you, Ma'am. I'll support you. It's so slippery.

> (*They keep moving forward, very cautiously –
> now, all three supporting one another.*)

ALISA. Do you think they have some dark plan in mind?

ANNA. God only knows. Many of the best workers have started drinking, just because their futures have been so unclear this past year.

MARTHA. So, they're drunk, too? Lord, help us, Alisa. What should we do? It was shameful for Karl to let us go out alone.

ALISA (*to Anna*). And *you*, of course, dump all the blame for this on the engineer?

ANNA. Well, of course, a simple worker can never understand how such things happen. I can only say that there are a lot of bad people among the workers, but none of them would get it into his head to marry someone else's wife. And we feel sure that such a marriage cannot have God's blessing.

ALISA. Anna, did you not love me before?

ANNA. How can you ask me that, Ma'am? But I don't love you, now… not as much as before. It's so hard to see how this could have happened to our sweet, lonely Alisa, whom we all loved so much. I remember how everyone adored you – both old and young. And everyone was sure that, one day, when Gerrgamra belonged to the young mistress, she would do much good for the people.

ALISA. Yes, but for now, I don't have the money to do anything, dear Anna.

ANNA. Oh, yes, I understand... But that you would agree to dismiss a number of the old workers... *That*, no one could ever have imagined.

ALISA. I don't want that.

MARTHA (*shouts*). Oh! Lord! Alisa! They're surrounding us! The entire mob!

(*Pressing in, the mass of workers surrounds them.*)

ALISA (*grabs Anna by the hand and tries to speak in a calm tone*). What does it mean? I don't want to talk to them. Not now! Let's go back! Get me to the lab!

ANNA (*as Erik is approaching Alisa*). There! There's my husband!

(*Martha shouts, rips the lantern out of Alisa's hands and runs back to the lab, continuing to cry out as she goes.*)

ERIK (*looks with scorn after her and then tells Alisa*). You did well to come. Here's the key to the battery. You protect him now, Ma'am, as you wish. I no longer can.

ANNA. Are they ready to beat you?

ALISA. Where's Ernest?

ERIK. He went to town. He probably went to summon the police. He ordered me to go to the shed and send a signal to the engineer, but the comrades won't allow it. You can try, Ma'am. Maybe you will succeed.

ALISA (*takes his key*). Of course.

(*She goes bravely to the shed.*)

VOICE. No! It's not going to happen! Give me that key!

(*Anders Gulte approaches her and wants to take away the key.*)

ALISA. Are you not ashamed? How can honest workers act like criminals and resort to violence?

SVEN KARLSON. Give me the key or we'll tear the shack to pieces!

OLD MAN. You, Ma'am, are to blame for everything. Why did you run away from your lawful husband and bring God's punishment on yourself and us?

OTHER. If her father had lived to see all the grief she's brought upon us – her father, who was always such a good master…

ALISA. Don't mention my father when you're acting with threats of violence. You would never dare do this to me if he were alive. And, if you still have a drop of attachment to his memory…

VOICE. Yes. Yes. God bless the late baron. Obeying him was a true pleasure.

(silence)

ALISA. You can see my fear. Don't continue to frighten me. Go home, quietly – for my sake!

ANDERS GULTE *(in a low-tone)*. If it were our baroness asking us, then it would be a different matter… but Torell's wife…

ALISA. So, now, you're reproaching me! Is there no one among you – you who have known me since childhood – no one who has kept the smallest shred of affection for me? *(silence)* Is there no one, not one, who remembers the child, Alisa, who ran about the workshops meeting affectionate smiles everywhere? Does no one have a kind word to say in my defense? *(silence)*

OLD MAN. Yes, you were a wonderful, captivating young girl. No one is going to deny that.

ALISA. Yes, Peter Ström. Both you and your wife have always loved me. Will you not speak up?

OLD MAN. Yes, I loved you before – when you were still innocent – but now that you have chosen a sinful life and have forgotten the laws of God and man…

ALISA. So, there's nobody, no one. Everyone here has turned away from me…

(Yalmar suddenly appears among them with a lantern in his hand.)

UNIVERSAL EXCLAMATIONS. Baron! Baron! Baron!

(Alisa retreats into the darkness, so that he doesn't see her.)

YALMAR. What's going on here? Can these be the same honest workers I know, now rebelling and making trouble! Have you changed so much during my year of absence?

SVEN KARLSON. Many things have changed at Gerrgamra since the baron left.

ANOTHER VOICE. But now we understand that you're returning, Baron.

YALMAR. It's still very doubtful. Don't expect it, and don't protest against engineer Torell. I have no doubt he will punish you brutally, if he becomes the master here.

VOICE. Is it possible that, in the end, he will be the owner?

YALMAR. Very possible.

SVEN KARLSON. But we don't want to have him as our master.

YALMAR. Is he so much unloved here?

VOICE. We all hate him.

YALMAR. That doesn't, actually, surprise me. He was always so sharp-tongued and harsh.

(Alisa approaches Yalmar.)

YALMAR. Alisa? You?! Here?!

ALISA. I'm here to defend my husband's rights against these rioters. And now you're inciting them even more.

YALMAR. I never expected this, Alisa – to see you in this position – in a hostile relationship to your beloved Gerrgamra workers.

ALISA. Protect me against this violence, Yalmar! Use their old respect for your authority. Order them not to interfere with me entering the shed and lighting the signal. This machine is my husband's most precious possession. It's his future – his everything.

YALMAR. You, yourself, I'm always ready to defend against any violence. But your husband should be defending his position here himself and offering them money. Come with me! I'm going to get you out of here! It's outrageous to see you here all alone. How could your husband allow this?

ALISA. If you don't want to help me notify him… *(She runs to the shed and puts the key in the lock, but it won't open. She puts her back to the door.)* I'm going to stand guard myself until you've all gone home – even if I have to stand here all night!

YALMAR. Oh! Fantastic show of enthusiasm! So, now you're going to play the role of defender of an industrialist in his unworthy struggle against the working people.

ALISA *(takes a step away from the door)*. My husband is not an industrialist. He's a scientist.

YALMAR. But the first use he makes of his scientific invention is to deprive fifty workers of their piece of bread.

ALISA *(moves even farther away from the door)*. It's not his fault. It's the fault of circumstance.

YALMAR. Do you think any circumstance could have made your father take such a step?

ALISA. It greatly distresses me.

YALMAR. But, he – a so-called friend of the workers – is ready for it. Follow my advice, Alisa, and at least don't accept personal involvement in this fight. Don't abandon the values you've grown up with. Come with me.

> *(He takes her hand. She allows herself to take a few more steps. This is immediately seen as an opening for Anders Gulte and Sven Karlson who rush towards the shed.)*

> *(Suddenly, Karl runs in without a hat, accompanied by Martha, who is hiding behind his coat. He charges for the shed, pushing past the workers and throws on the light. The whole scene is suddenly illuminated.)*

KARL. What kind of dark plots do you have going on here? I'm lighting up your filthy faces to find out who's here! *(He quickly comes towards Erik.)* You disobeyed orders and didn't send the signal!

Erik *(pointing to Alisa, who's still standing, completely stunned, underneath Yalmar's arm)*. I gave the key to the baroness.

KARL *(who only now notices Alisa and Yalmar)*. Alisa! You? Here?

ALISA (*rushes to him*). Karl! Talk to them. Tell them they won't be fired! Tell them you didn't create the invention for your sake only, but for their benefit as well.

KARL. Even when my life's work is in danger, you're on the side of my enemies instead of looking to defend me! So, I really am alone! They turned me away! They changed me! Well, good! Then so be it! In that case, I will defend myself!

> (*He rushes back towards the shed at the sight of
> Anders Gulte and Sven Karlson slipping into it.*)

KARL. Get out of there, you scoundrels, or I'll shoot you like dogs!

> (*He takes a revolver out of his pocket and directs it towards the open
> door of the shed. The two workers run out. The rest scatter in fright.*)

KARL. No matter the cost, I will defend my machine! You can be sure of that! It's worth much more than you scoundrels.

ANDERS GULTE. Here's a true friend of the people! Well, go ahead and shoot, villain! You'll be the worse for it. Here's Ernest with the police!

> (*Ernest is entering the scene with the police.
> The crowd quickly disperses*)

WORKERS GATHERED AT THE WATER WHEEL

Scene Created by Sandra DeLozier Coleman

ACT FIVE

The same scenery as in the last act. Early morning at sunrise. The workers are gathered again at the water wheel. Karl, seen in the shed where the door is open, is occupied with getting the machine to turn the wheel. Ernest and Erik are helping him.

STENSON (*enters*). So, everyone has gathered already. You're up early, my friends, considering how late you went home.

MASTER FOREMAN. If everyone slept as little as I did last night, then… But, what about you, Mister Director? Did you sleep here?

STENSON. Yes, I stayed at the hotel. It's too interesting. I couldn't leave. *(He points to Karl and Ernest.)* Have they been working all night?

MASTER FOREMAN. Yes. It seems something's not quite right.

STENSON (*laughing*). I would have thought so. At the last minute, something always goes wrong.

MASTER FOREMAN. Two workers were arrested – Anders Gulte and Sven Karlson. They're sitting in jail right now. The police took them in last night.

STENSON. What? Did the workers become violent?

MASTER FOREMAN. Yes. It was pitiful to see the engineer. He burst into tears like a child at the sight of the uprising – seeking to find out exactly who had had a hand in it.

STENSON. And what did his wife say?

MASTER FOREMAN. They say the crowd and commotion made her so ill last night that no one dared speak to her about damage to the machine. They sent for a doctor – afraid she would have an early delivery. The engineer, like a madman, was torn between her and the machine. He didn't seem to know which of them was more precious to him.

1ST WORKER. Yes, and, as he ran out of the house – where a few of us were standing outside – he yelled out at us: You've killed her, you bastards!

2ND WORKER. Yes, yes. We'll still have to pay for it, even if he manages to deal with the machine.

OLD MAN. But can it be possible that a tiny little thing in the shed could make this big wheel move? Do you believe it, Mister Director?

STENSON. No. To tell you the truth, my dear Peter Ström, I think, just like you, that all this is nothing more than madness. He will always be in need of some little spring or cog on which everything will depend. We have seen this, repeatedly.

SEVERAL VOICES (*shouting*). It's moving! It's moving!

1ST WORKER. It's a lie! It's the same as before.

2ND WORKER. If it fails, now, he will have to admit his overconfidence, since he has no money left to continue the experiments and there wouldn't be a single person left who would believe in him. Not even his wife.

1ST WORKER. Then he'd best be off to America.

MASTER FOREMAN. Oh! He will succeed! Rest assured! I've never seen fire in human eyes like the fire he had last night! Like a blasting hot furnace! I'm convinced that, even if he had to call on the devil himself to help, he wouldn't give up.

3RD WORKER. Then all of us who rebelled yesterday are going to be fired. And I'm going to have to move out of my home with my wife and six kids.

2ND WORKER. And I – with a wife who hasn't been out of bed for twelve years. Where would we go, now that it's so hard to find work anywhere?

1ST WORKER (*clenching his fist*). And, yet, people continue to invent these diabolical machines!

OLD MAN. It's so unfair – I'm sure the Lord will not allow it. I'll tell you something… I dreamt last night that the machine was torn apart as soon as the wheel began to move. It's a sign from the good Lord that he has heard my prayers.

3RD WORKER. It would be a good thing to be as faithful as old Peter. In every situation, he's able to find solace in higher things.

2ND WORKER. My sick wife is also always preaching about God's mercy. But we'll still have to starve, if we're chosen to be fired.

KARL (*comes out of the shed without a hat, with an exhausted face, pale and agitated. He shouts at two boys playing near the wheel.*) Get away from the wheel! It's starting to move!

(*The boys run away. There is movement among the people. All, with tense attention, watch the wheel. Karl rushes back into the shed.*)

1ST WORKER. It's moving as weakly as before. (*laughter*)

3RD WORKER. I'm sure he'll have to return to the old way.

STENSON. I think so, too, my friend.

(*At this moment, the wheel begins to move – first slowly, and then faster and faster, until, finally, it's in full swing. A few boys shout: Hurrah! They throw their caps in the air. There's a strong movement in the crowd. Karl, Ernest and Erik come out of the shed. Karl wipes the sweat off his face and falls, exhausted, onto a pile of firewood.*)

ERNEST (*runs onto the boardwalk, waving his cap and shouting*). Hey! Everybody! Let's hear it for the battery!

(*The cheers are piecemeal and unfriendly. Throughout the scene, women, children, workers, etc., move in from all sides, one after the other.*)

STENSON (*coming to Karl and squeezing his hand*). Didn't I always say it! You're a genius, my friend! I was always sure you'd be a celebrated person. You're just like your father – the wonderful old man, Torell. If only he had lived to see this!

(*Martha and Paula come running in.*)

ERNEST (*takes Martha by the hand*). It's good that you're here. Well... Let's get started, right now! (*He walks with her up to Stenson.*) Uncle Stenson!

STENSON. What's this?

ERNEST. You see... I played a part in this. I was always there, helping Karl.

STENSON. Well, what of it? (*He points to Martha.*) And she's here because?

ERNEST. Because she, also, played a role... because I was always in love with her, and when you love someone, you're more energetic... and you're able to work better when you have energy.

STENSON (*to Martha*). So, that's it! You let yourself fall in love without my knowledge.

MARTHA. No, Papa, dear. It wasn't me. It was Ernest…

STENSON (*laughing*). Ah! So, it's just Ernest – not you. Well, then there's nothing to say about it. I'm not going to force you to marry a man you don't love.

MARTHA. Come on, Papa! Shame on you!

(He stops his joking.)

STENSON. How cunning they are, these girls! Yes, dear Ernest, you will certainly get your share of the benefits, since you have helped so much.

KARL. Of course! We will share everything.

STENSON. That's splendid. Well, take him, daughter. You could do worse than to choose…

ERNEST. … a guy who's the brother of a great inventor!

MARTHA. … *and* his assistant! Don't forget that, because that's the most important thing! If you were only his brother…

ERNEST. You wouldn't want me…? Did you hear her…? How do you like that?!

(He kisses her.)

KARL (*runs towards Mdm Torell, who enters quickly from the left side*). So, what about Alisa? What's going on?

MDM TORELL. It's moving – really moving! Alisa assured me she'd seen from the window that the wheel was moving, but I didn't want to believe. God is merciful! It's moving!

(She falls on Karl's neck, crying.)

KARL. What is she saying? Is she awake? How's her health? What's she saying?

MDM TORELL. She cried. She became nervous, understandably – just as I did.

KARL. Is she dressed, yet? Ask her to come and see!

MDM TORELL. I don't think she'll want to come. She's so tearful. And she says she'll want to leave Gerrgamra soon.

KARL. To leave...? So, that's it! She's talking about that.

MDM TORELL. Yes, because she's so terribly saddened by the fact that the local people now have bitter feelings towards her. She could never stay here, she says. Oh! Karl! What will happen now? What a pity this is happening just now – just as you've finished and everything could be going so well!

KARL. Ask her to come here, so I can talk to her.

MDM TORELL. Why aren't you going to her?

KARL. I can't. I need what I have to say to be heard by others.

MDM TORELL. If that's how it is, okay, then... I'll tell her. *(She leaves.)*

(Yalmar can be seen far away at the side of the castle.)

VOICES *(among the workers)*. Look! There's the baron.

MARTHA *(squeezing Paula's hand)*. Do you see him?

PAULA. No... *(She hurries to meet him.)* I'm glad I have the chance to say goodbye to you.

YALMAR *(takes her hand and moves her aside)*. You want to leave, Paula? You can't be serious. I came here as fast as I could to keep you from going.

PAULA. I have no other choice. *(She withdraws her hand.)* I'll be back in a year after you will have married some noble girl.

YALMAR. You know perfectly well that I would never... I do wish to get back Gerrgamra – which I hope is a way to achieve personal happiness – and I am doing this solely out of a sense of self-preservation. You're right to despise me for that. I despise myself – but to live by giving music lessons, or as a wandering virtuoso... Don't you realize that I can't do that?

PAULA. Yes. That's why I'm leaving.

YALMAR. But I don't want you to leave. If you can't be my wife... Well... You could still be resolved to live here in Gerrgamra and be the mistress of the house and accept the neighbors. You're a free artist, and you love me. What stands between us?

PAULA (*retreating from him*). No, no, Yalmar. That would be humiliating!

YALMAR. Humiliating? Dear God! Do you think a woman's honor lies in a wedding ring and the title of *Madame*?

PAULA. No, that's not the point. I find it enlightening that there is nothing you want to sacrifice for me. You want to have all the blessings of life and me in the bargain. But I have more respect for myself than that.

> (*She runs away from him in dismay and goes to her mother, who appears on stage with Alisa. Yalmar paces back and forth, experiencing a strong inner struggle. Some of the workers begin approaching Alisa.*)

3RD WORKER. May I ask you, Ma'am, to put a word in to Mister Engineer for me? If I'm dismissed, my wife and children will be forced to live in a dependent's ward until I find a job, because, as you know, all my savings went to ashes during my illness.

2ND WORKER. And my poor sick wife can't be thrown out of bed.

KARL (*angrily, turning to them*). I forbid you to torment my wife with your stories. She's been ill and she can't be bothered. Anyone who dares approach her…

ALISA (*gives Karl an imploring look. She turns to the workers.*) I don't think we should stay here in Gerrgamra any longer, whatever the outcome of the settlement process. I can only wish you a good master, whoever he may be.

KARL. You want to leave your old home?

ALISA (*distressed*). My old home! The house I grew up in is no longer my home, if I have to live among my own workers as their enemy. To walk here and see all their empty houses and to be aware that the former inhabitants, whom I have known since childhood, were now just wandering – perhaps, homeless and begging. No, I couldn't bear that.

KARL (*calls to Mdm Torell, Yalmar, Ernest, Martha and Paula*). I want to say a few words to Alisa that you should all hear. (*speaking in an agitated voice*) This day should have been the best day of my life, but instead, it has been very different. Alisa is suffering a sorrow that she cannot overcome because she feels her former good relations with the people have been hopelessly damaged. I can't bear the idea that I'm the cause of so much unhappiness. So, I have made the following decision: If the settlement process is decided such that Gerrgamra should be sold, then I plan to employ all my income from the battery to purchase Gerrgamra in Alisa's name, and then she will have full freedom to realize her favorite dream – to give the factory to the workers association and live here as an equal with them.

ALISA. But Karl! What about you?

KARL (*smiling*). There is one condition – that you allow me enough money to continue with my scientific experiments.

ALISA. I don't understand what's happening here. When did you come up with this idea?

KARL. When? After such a night – after that awful moment when I saw you here, at this place. The idea entered my mind when I saw you standing there under Yalmar's arm, watching...

ALISA. Yes, it was an awful moment – standing here alone – seeing the hatred written on all the old familiar faces – and feeling that my faith in you was wavering.

KARL. And you had reason to waver, because you were right. I had pursued only my own interests. I realized that clearly the moment I found myself standing here with a revolver in my hand – aiming it at those who were here fighting for their daily bread.

ALISA. Yes, you holding a gun seemed to be an assault on everything I have loved in life.

KARL. Try to forget that image! I can't completely remake myself, but that's why I intend to give everything over to you.

ALISA. But it's not as though I want to be left alone – or for each of us to do our work separately. I've always dreamed of the two of us working *together*.

KARL. We'll work together if you can believe that all my inventions will benefit your new association. (*quietly*) And we'll keep our modest home!

ALISA. Yes. And the workers will gradually learn to understand and love you. They will see what broad and noble ideas you have.

KARL (*smiling*). Yes, sometimes a person can change for the better. (*whispering*) We can hope our baby will be born with better instincts.

MARTHA (*to Ernest*). I don't understand. Does Karl really have plans to arrange all this – including making a palace for the workers, and so on?

ERNEST. Oh, yes! Definitely.

MARTHA. So then, we'll do the same at Lido, Papa.

STENSON (*standing aside and calculating on paper*). What? I just made a calculation: We can reduce the working staff by at least 60 people.

MARTHA. No! What are you saying, Papa? We'll build a palace for the workers at the Lido. Yes, Papa, we must do it! Otherwise, Ernest and I are going to leave you. Really! We'll leave you. Isn't that right, Ernest?

STENSON. This craziness is contagious. That's the problem. But, fortunately, not everyone is infected. If Baron Yalmar wins…

YALMAR. Little hope for that… I prefer to voluntarily retreat. (*Paula gasps in delight.*) It's so difficult to go through this settlement process, and, besides, (*kissing Alisa's hand*) Gerrgamra belongs to you. When there are many contenders for one throne, the one with noble thoughts wins. I have no such thoughts, but I know how to respect them in others. (*to Karl*) In this respect, we're alike. So, we have an agreement. (*They shake hands, firmly. Then Yalmar takes Paula aside to talk to her.*)

ALISA. This is all so strange! I can scarcely believe it.

KARL. You can begin to carry out your plans right now. Go! Talk to the workers!

ALISA. Talk to them… I don't know what to say.

KARL. How can you not know what to say when you only need to tell them what you have desired and dreamed of all these years? The living room will turn into a classroom, the library into a living room…

ALISA (*smiling*). No, not into a living room – into a reading room. No, I'm too embarrassed to say it. It's so amazing – so unbelievable. It's such a big plan, so full of happiness that I can't even take it all in. To belong to each other – so completely, so fully – to share all our work, to aspire to one goal, to realize all our most beautiful dreams – dreams for ourselves and for others. No, it can't be true. My head is spinning.

(They hug, while Karl whispers something to her.)

PAULA *(goes to the podium arm in arm with Yalmar. She is speaking to him.)* We are now two composers writing a symphony about a waterfall. The sound is already in my ears. It's going to be a work of great genius.

YALMAR. Yes… Only I declare in advance the condition that, if there are curtain calls, I'll be spared the need to go out and bow to an audience of idiots. *(Kisses her hand.)*

MARTHA *(rushes to her father's neck)*. My sweet, sweet Papa – you'll see what a wonderful life this will be. And I'll arrange a school for needlework for young girls!

(There is noise among the workers at the sight of the return of the arrested Anders Gulte and Sven Karlson.)

ANDERS GULTE *(to some of the workers who surround him)*. The engineer sent an order to release us.

SVEN KARLSON. And we were told that we would retain our positions.

(There are sounds of approval among the workers.)

KARL *(runs up onto the podium and shouts)*. Workers! My wife wants to tell you something.

ALISA *(confused)*. No, Karl! How can I…?

(Karl helps her rise to the podium and leaves her there.)

ALISA *(tries to master her emotions, but, at first, cannot make a sound. Finally, she takes a deep breath and breaks the long silence with two words.)* My friends…

ANIUTA

EPILOGUE

Readers may wonder why Sofie and Anna Charlotte ended the plays with just the two words, *My friends...*! In her biography of Kovalevskaya, Anna Charlotte Leffler, says that the thoughts that were occupying Kovalevskaya's mind as she developed the plot and characters were chiefly thoughts of her sister, Aniuta.[70]

Since childhood, Sofie had had a strong attachment to Aniuta. In *A Russian Childhood*, she wrote that she "would have gone through fire and water" for her.[71] She recounts Aniuta's adolescent years, her struggles with accepting the inevitability of death, a period of religious asceticism, a phase of believing she was destined to become an actress, her absorption in the new social ideology, and the flowering of her passion to become a writer.

The heroine in Aniuta's first published story – the one she sent to Dostoevsky – is a well-bred young lady, like Aniuta herself, who meets a young student who impresses her deeply. She refrains from seeing the student again, solely because of social differences. In a dream, the student comes to her and shows her scenes from a life that *might have been* – a hard-working life with a person she loved, surrounded by intelligent friends – a life filled with warm, serene happiness and great hopes for the future.[72] Sofie says Aniuta pictured scenes like that for her own life and that she believed, deeply, in the possibility of that kind of happiness.[73]

Her father became ill when he discovered that his daughter had secretly become a published authoress. It was not until after many days of illness and anger that he had Aniuta read to him the story Dostoevsky had published. He recognized the sad heroine, so dissatisfied with her life, as his daughter, and gave Aniuta permission to correspond with Dostoevsky. Aniuta resumed her writing and became life-long friends with Dostoevsky, but still felt held back from her dreams of fame by her father's lingering opinion that writing was not a suitable occupation for a woman.

Aniuta was very seriously ill at the time Sofia was living in Stockholm, giving courses, doing research and writing. She and her husband, Victor Jaclard, were living in St. Petersburg when, for political reasons, Jaclard was ordered to leave the country. Aniuta had also been ordered to leave, but was too ill to travel unless transported on a stretcher. Sofia journeyed to Russia to stay with her sister during the last painful weeks of her life.[74]

At her sister's bedside, Sofia reflected on the critical, life-changing decisions she and Aniuta had made early in their lives.[75] Sofia had entered into a fictitious marriage that began with friendship and the shared ideal of making a positive difference in the world by opening the doors of education to women. She had believed that marrying Vladimir would serve this goal by making it possible for her to travel to where she could study and Aniuta could write without need for their father's permission.

Entering into a fictitious marriage was not unusual for young people caught up in the revolutionary thinking of the 1860's. They sincerely believed it was a noble and worthy sacrifice to forego personal satisfaction for the sake of providing a service to mankind in bettering society. In Sofia and Vladimir's case, the relationship was complicated. Vladimir did not simply drop his nominal wife off to study mathematics and then disappear from her life entirely, as did some nominal husbands. Sofie and Vladimir remained interested and involved in each other's lives and eventually grew close enough to decide to live as man and wife. They had one child, but never enjoyed a typical family life as they experimented with a number of consequential life choices.

Their attempts to live non-academic lives for a period of time and to make money by various business ventures failed miserably. Vladimir's distress over money problems and the shame of financial scandal eventually led him to take his own life. The tragic event undoubtedly influenced the ending in the play *How It Was* and led to serious reflection on how life might have been had they made other choices at critical moments in their lives.

Aniuta's struggle for happiness was also quite complicated. When Sofia married and invited her sister to live with her in Heidelberg, Aniuta did not stay with her sister for long. She soon moved to Paris, without either sister telling their parents. Aniuta had continued to dream of becoming a famous writer and had imagined, while in Heidelberg, that in Paris she might have opportunities to make progress toward her goal.

She was working as a typesetter,[76] when she met and became involved with Victor Jaclard, an active leader in the Paris Commune. Aniuta and Sofia had long been supporters of new ideas taking hold of people of their generation for building a society in which there would be greater equality among people and more wide-spread opportunity for education. Even at Palibino, Aniuta had impressed Sofia with her nobility of spirit by giving reading lessons to the children of the servants and engaging in long conversations with the peasant women.[77]

Aniuta became devoted to Jaclard's cause and to the efforts of the Commune to bring relief to the poorest of poor who were living in Paris under an oppressive government. She married Jaclard in a civil ceremony just a few days after he led a bloody battle for the people at Montmartre. She fully supported the cause of the common people in their rebellion against their oppressors and became one of the noted activists.[78]

Sofia joined Aniuta in Paris for 38 of the fearful 72 days of fighting. More out of love and concern for her sister than belief in the value of fighting, she helped Aniuta tend to the wounded. Both sisters were in danger of injury, death or arrest during that time. Vladimir had accompanied Sofie to Paris, but was not an active participant in the Commune. He spent his time there gathering information on paleontology from the Paris museums as bombs were exploding all around.[79] Books about the Commune generally include a beautiful portrait of Aniuta, but the books I bought in Paris do not include Sofia, Vladimir, or Jaclard.

Aniuta, like the character Alisa in the plays, had been born to aristocracy and comforts, but in Paris she lived and worked among the impoverished people she was hoping to help. She saw them as friends in need of her friendship and support. She could have been the model for the character Alisa, who at the end of the play wants to share all her wealth with the factory workers through a plan for converting Gerrgamra into a school and gathering place for the workers, who she planned to treat as her friends.

Kovalevskaya, herself, could also be a model for Alisa. The one time she and Vladimir did experience a temporary period of financial success, she proposed a plan similar to Alisa's for helping poor people. She dreamed of an idealized society of the future, in which the happiness of one would be the happiness of all; the sufferings of one, the sufferings of all.[80]

Although Sofia's idea of how best to move toward a more just and equal society did not take the form of active rebellion, she admired her sister's courage, even as she feared for her safety. The battles fought during the Paris Commune accomplished none of the goals of the Communards and ended in the death or deportation of thousands. After the battle was lost, Aniuta and Jaclard barely managed to escape to Switzerland, but they were never really free from the repercussions of having participated in the fighting, even after the Communards were given amnesty. That loss of freedom was a great sacrifice and affected the rest of Aniuta's life.

As her sister suffered through her long illness, Sofie may have hoped that what she was writing would show Aniuta, and the rest of the world, that she viewed her sister's desire to befriend people less fortunate than herself as a mark of true nobility – a higher form of nobility than the documentation of noble heritage their father had worked so hard to obtain when they were children. [81] Yalmar expresses this idea at the end of *How It Might Have Been,* when he explains why he feels Alisa deserves to inherit Gerrgamra.

It was so important to Kovalevskaya to write the two plays and to get them published quickly that she put her work on the Prix Bordin problem aside to focus on the writing. She hurried Anna Charlotte,[82] who wrote sketches of the first five acts in five days! Sofie probably hoped to finish while her sister was still alive. She may have thought Aniuta might find peace in a positive portrayal of the noble motivations behind the difficult, consequential, and much-criticized choices Aniuta had made in her difficult life.

With all this considered, the words, *My friends...,* become powerful words of hope and beginnings, rather than an end. We are expected to continue Alisa's speech ourselves, as we recall the words of her idealized plans for a workers' palace. *The big hall will be turned into a school – the billiard room, into a lecture hall – the library, into a reading room...* [83]

APPENDIX A

1842 Future husband, Vladimir Kovalevsky, a respected paleontologist who worked with Darwin, was born on August 2.

1843 Future parents were married on January 17th. Her sister, Anna Vasilevna Krukovskaya, was born on October 6.

1850 Sofia Vasilevna Krukovskaya (SK) was born on January 15 in Moscow.

1852 Father, Vasily Vasilevich Krukovsky, received his generalship.

1855 A brother, Feodor (Fedya), was born. The SK family moved on military orders to Kaluga, where they lived for three years.

1858 An English governess, Margarita Smith, replaced the French governess and SK's beloved Nanny. The family began living at Palibino, the large country estate SK's father had inherited in 1843. Calculus notes were substituted for wallpaper in her room. This was the year the family was given the right to refer to themselves as Korvin-Krukovsky, signifying their descent from a Hungarian king.

1864 SK began teaching herself Trigonometry.

1868 SK entered into a fictitious marriage with Vladimir Onufrievich Kovalevsky (VK) in September. The marriage allowed SK to obtain a passport to travel in search of a university that would allow her to attend classes. As a married woman, she would be able to host relatives and women friends who were also desiring higher education outside Russia.

1869 Sofia and Vladimir left St. Petersburg in April for Vienna, where SK was unable to find math professors willing to accept her as a student. SK began attending the University of Heidelberg in Germany in the fall. She accompanied Vladimir on a trip to London during the October break, where she met Thomas Huxley and Charles Darwin and discussed with Herbert Spencer "woman's capacity for abstract thought."

1870 SK arrived in Berlin in October hoping to study under Karl Weierstrass.

1871 SK Participated in the Paris Commune through hospital work from April 5 through May 12 with sister, Anna, and Anna's future husband, Victor Jaclard. Vladimir explored the Paris paleontological resources there. When Jaclard was arrested, SK's father came to the rescue and exercised influence to free him.

1873 Anna and Victor Jaclard's son Urey was born.

1874 SK prepared three papers for a doctoral dissertation. She was awarded a Ph.D. in Mathematics summa cum laude in absentia by the University of Göttingen. She became the first woman to be granted a doctoral degree in mathematics. Sofia and Vladimir returned for a while to the Palibino estate in Russia in mid-August.

1875 Father, General Vasily Korvin-Krukovsky died on September 30th of an aneurism.

1876 Sofia and Vladimir began to live as husband and wife.

1878 Daughter, Sofia (Fufa), was born October 17.

1879 Mother, Elizaveta Feodorovna Korvin-Krukovskaya (Shubert), died.

1881 SK separated from her husband, Vladimir.

1883 Vladimir committed suicide in despair over failed business dealings. SK accepted an unpaid position as a privat-docent at Stockholm University in Sweden.

1884 SK was appointed to a five-year position at Stockholm as Professor. She became the editor of Acta Mathematica.

1887 Sister, Aniuta (Anna Jaclard), died suddenly while Sofia was working on the paper for the Prix Bordin and *The Struggle for Happiness*. Sofia had been at her bedside earlier in the year in Russia.

1888 In February, Maxime Kovalevsky, a Russian professor who was well-known for his writings and lectures on the social sciences, arrived in Stockholm. In December, SK was awarded the French Academy of Science Prix Bordin for her solution of a long-standing problem related to the revolution of a body about a fixed point.

1889 SK was awarded a life-time full professorship at the University of Stockholm.

1889 She published her childhood memories in Swedish under the title *The Rajevski Sisters*.

1890 Kovalevskaya published her childhood memories in Russian under the title *A Russian Childhood*.

1891 SK died on February 10 in Stockholm of pneumonia after a trip to see Maxime in Beaulieu-Sur-Mer, France.

APPENDIX B
RHYMED VERSION OF CHAMELEON

This version of *Chameleon*, which was translated by A. De Furuhjelm and A. M. Clive Bayley, appears in Anna Charlotte Leffler's biography of Sofia Kovalevskaya, written shortly after her death.

Chameleon

« Kameleonten känner du från barndoms dagar.
Så snart han sitter ensam i sin vrå
Är han så anspråkslös, så ful och grå,
Men sedd vid rätta ljus kan också han bli fager.

"The changeful chameleon as everyone knows,
As long as he sits alone in his nook,
Is ugly and dull and grey in his look;
But in a good light how brightly he glows.

« Han ingen skönhet har, han återspeglar bara
Allt hvad han ser omkring sig, skönt och godt.
Han skimra kan i gult, i grönt, i blått,
Så som hans vänner är, så skall han också vara.

"No beauty has he, but he always reflects
What around him exists of beautiful hue.
He can shimmer alike in gold, green, or blue
And of all his friends' hues there is none he rejects.

« I detta djur mig syns jag ser min egen like,
Min kära vän, hvarthän du också går,
Jag följer alltid troget dina spår;
Jag står ej efter, jag, jag ger väl aldrig vike.

"In this creature, me seems, my likeness I see,
For, dearest of friends, wherever you go
I go in your steps; for it is aye so,
That I can't stand behind, nor be turned back from thee.

« Har man en vän som du, för äran får man plikta,
Du skrifver, målar, ritar etcetera.
För mig är detta småsak, patrata.
Men Gud förbarme sig! Nu vill du också dikta ! »

To a friend such as you all my reverence is due,
You write and you paint and you draw and whatnot.
The things are to me but rubbish and rot,
But, oh mercy on me! You poetize too!"

APPENDIX C
RHYME SCHEME FOR ГРУНЯ (GRUNYA)

1.	Готовится святым страдальческий венец.	a
2.	И сердце бьется в ней при мысли, что, быть может,	b
3.	И ей такой на долю выпадет конец.	a
4.	Повсюду мысль ее преследует и гложет,	b
5.	Как пальмы вечные сподвижников святых —	c
6.	Себе стяжать; нередко пылкое мечтанье	d
7.	Рисует ей картину мук и пыток злых.	c
8.	По вечерам в лучах заката огневых	c
9.	Ей грезится костра багровое сиянье.	d
10.	Ей слышен гул колоколов, цепей бряцанье,	d
11.	Оружья звон; народом площадь вся полна.	e
12.	Ее ведут на казнь, спокойна и ясна.	e
13.	Восходит Груня на костер; в руках распятье	f
14.	Одежды мрачные; ни стона, ни проклятья,	f
15.	Господний дух ее невидимо крепит.	g
16.	Господний дух речет ее устами.	h
17.	Пред казнею к народу Груня говорит.	g
18.	И, потрясен ее могучими словами,	h
19.	Народ в безмолвии рыдает и дрожит.	g
20.	Толпа покаяться, уверовать готова,	i
21.	Но вспыхнул ярко вдруг костер, и дым багровый	i
22.	Обвился вкруг нее; мучений краткий миг,	r
23.	Стенанье слабое, один предсмертный крик,	r
24.	И кончено уж все, и длинной вереницей	t
25.	Летят уж ангелы за чистою душой,	s
26.	Кругом нее все свет, все ангельские лица,	s
27.	Навстречу ей пахнуло райскою струей.	t
28.	Уж рай пред ней святые двери отворяет…	u
29.	Но вдруг рассеялась мечта, и Груня замечает,	u
30.	Что это все один пустой, горячий бред.	v
31.	Кругом все прежнее; на небе догорает	u
32.	Заката яркого последний бледный след.	v
33.	И долго Груня все с поникшей головою	w
34.	Сидит перед окном печальна и грустна.	x
35.	И ночь идет с своею мрачной пеленою,	w
36.	Над кладбищем восходит яркая луна.	x
37.	Могилы и кресты рисуются так живо	y
38.	В ее причудливых, серебряных лучах.	z
39.	Стрекозы шумные трещат в траве болтливо,	y
40.	И соловей поет в сиреневых кустах.	z

APPENDIX D
BOOKS REVIEWED FOR THE AMATYC REVIEW

A CONVERGENCE OF LIVES, Ann Hibner Koblitz, Rutgers University Press, 1993, xxxviii + 305 pages, ISBN 0-8135-1962-4 / 0-8135-1963-2 pbk.

A RUSSIAN CHILDHOOD, Sofya Kovalevskaya, translated by Beatrice Stillman, Springer-Verlag, 1978, xiii + 250 pages, ISBN 0-387-90348-8.

LITTLE SPARROW, Don H. Kennedy, Ohio University Press, 1983, ix + 341 pages, ISBN 0-8214-0692-2 / 0-8214-0703-1 pbk.

INFINITY AND THE MIND The Science and Philosophy of the Infinite, Rudy Rucker, Princeton University Press, 1995, xii + 342 pages, 107 figs., 3 tables, ISBN 0-691-00172-3 pbk.

IN SEARCH OF INFINITY, N. Ya. Vilenkin, translated by Abe Shenitzer with the editorial assistance of Hardy Grant and Stefan Mykytiuk, Birkhauser/Boston, 1995, vii + 145 pages, ISBN 0-8176-3819-9/ISBN 3-7643-3819-9.

FLATLAND A Romance of Many Dimensions, Edwin A. Abbott, A Signet Classic, Penguin Books USA Inc., 1984, 160 pages, ISBN 0-451-52290-7 pbk.

ALICE IN QUANTUMLAND An Allegory of Quantum Physics, Robert Gilmore, Copernicus, an imprint of Springer-Verlag New York, Inc., 1995, ix + 184 pages, ISBN 0-387-91495-1.

POETRY OF THE UNIVERSE, A Mathematical Exploration of the Cosmos, Robert Osserman, Anchor Books-Doubleday, 1995, xiv + 210 pages, ISBN 0-385-47340-0.

THE ILLUSTRATED A BRIEF HISTORY OF TIME, Updated and Expanded Edition, Stephen Hawking, Bantam Books, 1996, iv + 248 pages, ISBN 0-553-10374-1.

HELAMAN FERGUSON: Mathematics in Stone and Bronze, text by Claire Ferguson, Meridian Creative Group, 1994, xiii + 79 pages, ISBN 0-9639121-0-0.

THE JUNGLES OF RANDOMNESS: A Mathematical Safari, by Ivars Peterson, John Wiley& Sons, Inc., 1998, xiii + 239 pages, ISBN 0-471-16449-6.

THE EMPEROR'S NEW MIND: Concerning Computers, Minds, and the Laws of Physics, by Roger Penrose, Oxford University Press, 1989, xiii + 460 pages, ISBN 0-19-851973-7.

THE CAMBRIDGE QUINTET: A Work of Scientific Speculation, by John L. Casti, Helix Books/ Addison Wesley, 1998, xxiii + 182 pages, ISBN 0-201-32828-3.

MATHEMATICIANS ARE PEOPLE, TOO, Stories from the **Lives of Great Mathematicians**, Luetta Reimer and Wilbert Reimer, D. Seymour Publications, 1990, viii + 143 pages, ISBN 0-86651-509-7.

MATHEMATICIANS ARE PEOPLE, TOO, Stories from the **Lives of Great Mathematicians**, Volume Two, Luetta Reimer and Wilbert Reimer, Dale Seymour Publications, 1995, viii + 144 pages, ISBN 0-86651-823-1.

THE PÓLYA PICTURE ALBUM: Encounters of a Mathematician, by George Pólya, edited by G. L. Alexanderson, Birkhäuser, 1987, 160 pages, ISBN 3-7643-3352-9

MATHEMATICAL ENCOUNTERS of the Second Kind, Philip J. Davis, Birkhäuser Boston, 1997, 304 + viii pages, ISBN 0-8176-3939-X.

THE MAN WHO LOVED ONLY NUMBERS: The Story of Paul Erdös and the Search for Mathematical Truth, Paul Hoffman, Hyperion, New York, 1998, 284 pages plus notes and index, ISBN 0-7868-8406-1.

ZERO: The Biography of a Dangerous Idea, Charles Seife, Viking, New York, 2000, 248 pages, ISBN 0-670-88457-X.

FERMAT'S LAST THEOREM: Unlocking the Secret of an Ancient Mathematical Problem, Amir D. Aczel, Dell Publishing, New York, 1996, 142 pages + index , ISBN 0-385-31946-0.

UNCLE PETROS & GOLDBACH'S CONJECTURE, Apostolos Doxiadis, Bloomsbury USA, New York, 2000, 209 pages, ISBN 1-58234-067-6.

EUCLID'S WINDOW, The Story of Geometry from Parallel Lines to Hyperspace, Leonard Mlodinow, The Free Press, New York, 2001, ISBN 0-684-86523-8.

A BEAUTIFUL MIND: A Biography of John Forbes Nash, Jr., Winner of the Nobel Prize in Economics, 1994, Sylvia Nasar, A TOUCHSTONE Book, published by Simon & Schuster, New York, 1998, 459 pages, ISBN 0-684-85370-1.

THE ESSENTIAL JOHN NASH, edited by Harold W. Kuhn and Sylvia Nasar, Princeton University Press, Princeton and Oxford, 2002, 269 pages, ISBN 0-691-09527-2

EINSTEIN IN LOVE, A Scientific Romance, Dennis Overbye, VIKING, Penguin Putnam, Inc., New York, 2000, 416 pages, ISBN 0-670-89430-3.

THE EXPANDED QUOTABLE EINSTEIN, collected and edited by Alice Calaprice, Princeton University Press, Princeton and Oxford, 2000, 407 + xliii pages, ISBN 0-691-07021-0.

EINSTEIN'S DREAMS, A Novel, Alan Lightman, Warner Books, A Time Warner Company, New York, 1993, 179 pages, ISBN 0-446-67011-1.

BEYOND THE LIMIT, The Dream of Sofya Kovalevskaya, Joan Spicci, A Forge Book, Tom Doherty Associates, LLC, New York, 2002, ISBN 0-765-30233-0.

NIHILIST GIRL, Sofya Kovalevskaya, translated by Natasha Kolchevska with Mary Zirin, The Modern Language Association, New York, 2001 ISBN 0-87352-790-9.

FRAGMENTS OF INFINITY, A Kaleidoscope of Math and Art, Ivars Peterson, John Wiley & Sons, Inc., New York, 2001, ISBN 0-471-16558-1

777 MATHEMATICAL CONVERSATION STARTERS, John dePillis, Spectrum Series, The Mathematical Association of America, Inc., United States of America, 2002, ISBN 0-88385-540-2

THE GOLDEN RATIO: THE STORY OF PHI, the World's Most Astonishing Number, Mario Livio, Broadway Books, New York, 2002, ISBN 0-7679-0816-3.

THE MYSTERY of THE ALEPH, Mathematics, the Kabbalah, and the Search for Infinity, Amir D. Aczel, Washington Square Press publication of POCKET BOOKS, a division of Simon and Schuster, Inc., New York, 2000, ISBN 0-7434-2299-6.

FRACTALS, GRAPHICS, AND MATHEMATICS EDUCATION, Michael Frame and Benoit B. Mandelbrot, editors, The Mathematical Association of America., 2002, ISBN 0-88385-169-5.

FROM ZERO TO INFINITY: What Makes Numbers Interesting, 50th Anniversary Edition, Constance Reid, A. K. Peters, Ltd., Wellesley, Massachusetts, 2006, ISBN 1-56881-273-6.

THE CALCULUS WARS: Newton, Leibniz, and the Greatest Mathematical Clash of All Time, Jason Socrates Bardi, Thunder's Mouth Press, an imprint of Avalon Publishing Group, Inc., New York, 2006, ISBN 1-56025-706-7.

TOM STOPPARD: PLAYS 5 – *Arcadia, The Real Thing, Night and Day, Indian Ink, Hapgood,* Tom Stoppard, Faber and Faber Limited, London, 1999, ISBN 0-571-19751-5.

THE PARROT'S THEOREM: A Novel, Denis Guedj, Translated by Frank Wynne, Thomas Dunne Books, an imprint of St. Martin's Press, New York, 2000, ISBN 0-312-30302-5 pbk.

CRIMES AND MATHDEMEANORS, Leith Hathout, Illustrated by Karl H. Hofmann, A.K. Peters, Ltd., Wellesley, Massachusetts, ISBN-10: 1-56881-260-4.

LETTERS TO A YOUNG MATHEMATICIAN, Ian Stewart, Basic Books (a Member of the Perseus Books Group), New York, NY, 2007, ISBN: 9780465082322, ISBN-10: 0-465-08232-7 pbk.

YEARNING FOR THE IMPOSSIBLE – The Surprising Truths of Mathematics, John Stillwell. A.K. Peters, Ltd., Wellesley, Mass, 2006. Hardcover, 244 pp. ISBN 978-1-56881-254-0.

APPENDIX E

Жон 3:16
Ибо так возлюбил Бог мир, что отдал
Сына Своего единородного, дабы
всякий, верующий в Него, не погиб,
но имел жизнь вечную.

Drawing by the author's sister, Sharon Woodman,
spoken of in the chapter *Searching for the Soul of a Poet*[84]

BIBLIOGRAPHY

A Convergence of Lives, Ann Hibner Koblitz, Rutgers University Press, 1993, xxxviii + 305 pages, ISBN 0-8135-1962-4 / 0-8135-1963-2 pbk.

A Russian Childhood, Sofya Kovalevskaya, translated by Beatrice Stillman, Springer-Verlag, 1978, xiii + 250 pages, ISBN 0-387-90348-8.

Beyond the Limit, The Dream of Sofya Kovalevskaya, Joan Spicci, A Forge Book, Tom Doherty Associates, LLC, New York, 2002, ISBN 0-765-30233-0.

Briefwechsel zwischen Karl Weierstraß und Sofija Kowalevskaja, Reinhard Bölling, Berlin Akademie Verlag,1993, 504 pages, ISBN 3-05-501338-7.

Das Fotoalbum für Weierstraß / A Photo Album for Weierstrass by Reinhard_Bölling, Vieweg & SohnVerlagsgesellschaft mbH, Braunschweig/Wiesbaden, 1994, ISBN 3-528-06602-4.

Dostoevsky Reminiscences, Anna Dostoevsky, translated and edited by Beatrice Stillman with an introduction by Helen Muchnic, Liveright, NY, 1975, 449 pages, ISBN 13: 978-0-87140-117-5.

Gösta Mittag-Leffler, A Man of Conviction, Arild Stubhaug, translated by Tiina Nunnally, Springer-Verlag Berlin Heidelberg, 2010, 734 pages, ISBN 978-3-642-11671-1.

Helaman Ferguson: Mathematics in Stone and Bronze, text by Claire Ferguson, Meridian Creative Group, 1994, xiii + 79 pps, ISBN 0-9639121-0-0.

Iconographie de la Commune de Paris de 1871, Gérald Dittmar, Éditions Dittmar, 2005, 510 pages, ISBN 9782916294018.

La Commune(Paris, 1971), A Film by Peter Watkins, ICARUS Films, 2000, ISBN 854565001213, DVD.

Little Sparrow, Don H. Kennedy, Ohio University Press, 1983, ix + 341 pages, ISBN 0-8214-0692-2 / 0-8214-0703-1 pbk.

Love and Mathematics, Pelageya Kochina (P. Ia. Polubarinova-Kochina), translated by Michael Burov, Mir Publishers Moscow, 1985, 340 pages, ISBN 978-0-82853-373-7.

Mon Choix de Sophie, Michèle Audin, Institute de Recherche Mathématique Avancée, Université Louis Pasteur et Le Centre National de la Recherche Scientifique, 2006, 25 pages, http://irma.math.unistra.fr/~maudin/choix-sophie.pdf

Nihilist Girl, (Sofya Kovalevskaya), translated by Natasha Kolchevska with Mary Zirin, The Modern Language Association of America, New York, 2001, 139 pages, ISBN 0-87352-790-9.

Remembering Sofya Kovalevskaya, Michèle Audin, Springer-Verlag, London, 2011, 284 pages, ISBN 978-0-85729-929-1.

Sonya Kovalevskaya, a Biography written by Anna Carlotta Leffler, Duchess of Cajanello, and *Sisters Rajevsky, Being An Account Of Her Life,* originally written in Russian by Sonya Kovalevsky and translated into Swedish by Anna Carlotta Leffler, translated from Swedish into English by A. De Furuhjelm and A.M. Clive Bayley, with a biographical note by Lily Wolffsohn, T. Fisher Unwin, Paternoster Square, London, 1895. 177 + 200 pages, ISBN 978-1-333-54279-5.

Une tragédie française : La Commune de Paris de 1871, Gérald Dittmar, Éditions Dittmar, 2006, 397 pages, ISBN 978-2-91629-411-7.

Vospominaniya Povesti, (the Russian language collection of Sofia Kovalevskaya's writings), edited by P. Ia. Polubarinova-Kochina, Nauka: Moscow, 1974, 561 pages.

NOTES

1 Isaak Brodsky's painting of workers at the Putilov factory

File:Isaak Brodsky putilov.jpg. (2020, June 23). Now in public domain. *Wikimedia Commons, the free media repository.*

https://commons.wikimedia.org/w/index.php?title=File:Isaak_Brodsky_putilov.jpg&oldid=428410481.

http://www.nepalenergyforum.com/hydropower-means-a-lot-worldwide/

2 I served as Book Review Editor for *The AMATYC Review* from 1996 to 2008. A list of books reviewed is in APPENDIX D.

3 *Beyond the Limit,* Joan Spicci, published in 2002 by Tom Doherty Associates, New York was another Sofia Kovalevskaya book I wrote about in *The AMATYC Review.*

4 As I reread my words, I made a few corrections and edits to the previously published review.

5 Those who may be interested in learning more about Sofia Kovalevskaya, should be aware that the translation of a Russian name into an English or German equivalent is not always consistent. Expect Kovalevskaya's name to appear in other forms, such as Kovalevskaia, Kovalevsky, or Kowalevskaya and to find her referred to variously as Sofie, Sofia, Sofya, Sophie, Sonya, and Sofa, due to both variations in transliteration and level of intimacy.

6 There is no Nobel Prize in Mathematics. Stories abound as to the reason. In any case, the French Academy of Science Prix Bordin is a prize of that level of significance. Submissions were presented in sealed envelopes on which each submitter wrote a short maxim or proverb. Sophia Kovalevskaya wrote "Dis ce que tu sais, fait ce que dois, advienne que pourra." (Say what you know, do what you must, come what may) *Mon Choix de Sophie,* Michèle Audin, Mar 2006, *p 19.*

http://irma.math.unistra.fr/~maudin/choix-sophie.pdf

7 Aleksandra Stanislavovna Montvid was a Ukrainian-born Russian short-story writer who wrote under the pseudonym, A. S. Shabelskaia. The words conveyed by Kovalevskaya in the response to the letter she received appear in Don H. Kennedy's *Little Sparrow,* pp 263 – 264. The original fan letter is at the Mittag-Leffler Institute.

8 The original text was written in Russian, but was first published in Swedish under the title *From Russian Life: Sisters Rajevsky*. Anna Charlotte Leffler includes the Swedish version of the memoirs in one of the publications of the Kovalevskaya biography she wrote shortly after her friend's death.

9 TODAY IN SCIENCE HISTORY: Science Quotes by Sofia Kovalevskaya

https://todayinsci.com/K/Kovalevskaya_Sofia/KovalevskayaSofia-Quotations.htm

10 Koblitz, *A Convergence of Lives*, p 257.

11 Don H. Kennedy, *Little Sparrow*, p 263.

12 The St. Petersburg symposium in May of 2000 on *The Theory of Partial Differential Equations and Special Topics on the Theory of Ordinary Differential Equations dedicated to the 150th Anniversary of the Birth of Sofia Kovalevskaya* attracted mathematicians from many countries, including Israel, Sweden, Belgium, France, Italy, Switzerland, the United States and Germany, as well as Russia.

13 See the reviews included in the PREFACE.

14 My love for writing brought many penpals into my life. My most treasured penpal was a young man who lived in Kagoshima-Ken, Japan. Takami Tanoue wrote very enthusiastically about how much he enjoyed my letters and my way of writing. He said friends, who were also part of his school's penpal exchange, envied his correspondence with me. I still have the letters, photos and gifts he sent. A fan with a painting of a silkworm on it was 100 years old when he sent it and is a special treasure still hanging on a wall in my home. It is historically interesting that every time I tried to find something very *American* to send to him, like a small, porcelain, Native American doll, I discovered a label marked Made in Japan!

15 I was chosen for *Quill and Scroll* because of my writing and because I served as Front Page Editor for our school newspaper.

16 The American Mathematical Society and the Mathematical Association of America.

17 I completed the translation of *Grunya*, the last and most difficult of the poems, in July of 2021. In the appendices, I have provided the Russian version from *VP* to show the complexity of the rhymes used in the rhymed version. Matching this rhyme scheme without loss of meaning was not easy.

18 This is a link to one of the programs for *An Evening of Poetry & Art* sponsored by SIGMAA-ARTS and the Journal of Humanistic Mathematics in which I was an invited reader.

https://scholarship.claremont.edu/jhm/poetry_reading_2016.pdf

SIGMAA-ARTS is the Special Interest Group of the MAA on Mathematics and the Arts. The group supports and encourages the interest of the mathematics community in the connections between mathematics and the arts including, but not limited to, the visual arts, sculpture, architecture, origami, textile and fiber art, literature, drama, dance, music, multimedia art, and digital art.

http://sigmaa.maa.org/arts/

The Journal of Humanistic Mathematics (JHM) explores mathematics as a human endeavor in the tradition of the Humanistic Mathematics Network Journal founded by Alvin White. JHM publishes mathematical poetry, fiction, personal essays, research studies, and more.

http://scholarship.claremont.edu/jhm/

19 *Helaman Ferguson: Mathematics in Stone and Bronze*, text by Claire Ferguson, Meridian Creative Group, 1994, xiii + 79 pps, ISBN 0-9639121-0-0.

20 The following is a link to the *Roots of Unity* article published by Evelyn Lamb on January 13, 2013 entitled *Setting Mathematics in Verse*. Evelyn Lamb also used the art I had submitted to the conference art gallery to illustrate the article and then used it once again in a later article entitled *Can You Tell the Difference between Math and Poetry?*
https://blogs.scientificamerican.com/roots-of-unity/setting-mathematics-in-verse/
https://blogs.scientificamerican.com/roots-of-unity/mathematics-or-poetry-take-the-quiz/

21 JoAnne Growney's blog, *Intersections–Poems with Mathematics*.
https://poetrywithmathematics.blogspot.com/search/label/Sandra%20DeLozier%20Coleman

22 Pierre Carrée: Des maths *(mais pas seulement) pour mes élèves (et les autres)*.
https://clairelommeblog.wordpress.com/2016/08/16/poesies-mathematiques/

23 The symposium on *The Theory of Partial Differential Equations and Special Topics on the Theory of Ordinary Differential Equations dedicated to the 150th anniversary of the Birth of Sofia Kovalevskaya*. Photos are available here:
www.pdmi.ras.ru/EIMI/2000/sofia/
www.pdmi.ras.ru/EIMI/2000/sofia/alb.html

24 A very interesting brief biography of Olga Alexandrovna Ladyzhenskaya (1922-2004) can be found at the link below. She also was interested in poetry and the plight of poets in her youth.

https://www.rbth.com/history/330424-female-scientists

25 The link below takes the viewer on a wonderful tour of the Institute Mittag-Leffler with stories of Sofia Kovalevskaya included. This is still the repository for many significant original documents related to the Kovalevskaya story.

https://www.youtube.com/watch?v=gShm4n999kM

26 The Russian text of *Vospominaniya Povesti* is now available on-line.

https://imwerden.de/pdf/kovalevskaya_vospominaniya_povesti_1974_text.pdf

27 Thankfully, over the years, tools for scanning and translating, as well as my personal Russian language skills, were improving significantly. I still cannot think of myself as a person who speaks Russian, however. My writing skills have made good translation possible with the help of translation software, but only immersion will make me able to actually speak Russian well and it is unlikely that I will ever have the opportunity to converse on a daily basis in Russian.

28 I can imagine readings without costumes or stage settings in lieu of full productions. I have attended effective readings, where the characters were played by un-costumed readers sitting in folding chairs. This was easy to accomplish for *Waiting for Godot*, but would probably require some sort of program with character names associated with each chair for these plays. I can also imagine a film based on the second play.

29 Kennedy, *Little Sparrow,* p 268.

30 Koblitz, *A Convergence of Lives,* pp 201 – 202.

31 Stubhaug, *Gösta Mittag-Leffler, A Man of Conviction*, pp 378 – 379.

32 Bölling, *Briefwechsel zwischen Karl Weierstraß und Sofija Kowalevskaja,* pp 290 – 293. Letter dated August 27, 1883.

33 Maxim Maximovitch Kovalevsky has the same last name as Sofia Kovalevskaya, because he was a distant cousin of her husband, the paleontologist Vladimir Onufrievich Kovalevsky. Maxim Kovalevsky and Sofia Kovalevskaya were never married. The identification of Maxim Kovalevsky as Sofia Kovalevskaya's fiancé is complicated by conflicting accounts of their relationship. It is certain that she was in love with him. They had certainly discussed marriage and may have agreed to marry just before Kovalevskaya's death in February of 1891.

34 The biography of P. Ia. Polubarinova-Kochina found at the link below will amaze readers. Beyond making astounding contributions to mathematics, most notably in the areas of fluid mechanics and hydrodynamics, her writings have added significantly to the annals of math history. It was she who published the revision of *Vospominaniya Povesti* from which I translated these plays and poems. She wrote several books about Sofia Kovalevskaya, including *Love and Mathematics: Sofia Kovalevskaya*, and published biographies of many other significant figures in the mathematics community, including Weierstrass, Mittag-Leffler and her husband, Nikolai Yevgrafovich Kochin. Much of her writing was done after the age of 86. She lived until 1999, her 100th year – the year in which she published her final paper on applications to problems of underground hydromechanics.

https://mathshistory.st-andrews.ac.uk/Biographies/Kochina/

35 Sofia Vladimirovna Kovalevskaya (Fufa) (1878-1952) never married. Ann Hibner Koblitz in *A Convergence of Lives* gives a few details about Fufa's life in a long note on page 236.

36 Sofya Kovalevskaya, *A Russian Childhood.*

37 See Appendix for rhymed version of *Chameleon.*

38 Kochina, *Love and Mathematics*, p 247.

39 Koblitz, *A Convergence of Lives,* p 159.

40 Vladimir Onufrievich Kovalevsky was a paleontologist who worked with Darwin and produced Russian translations of Darwin's work. Darwin expressed great respect for Kovalevsky's work in evolutionary science. Readers familiar with a chart illustrating the evolution of the horse have seen results of his work. For at least five years Vladimir and Sofia continued in a nominal marriage. After they began to live as man and wife, the marriage produced one child, a daughter Kovalevskaya named after herself, but called Fufa or Fifi.

41 Koblitz notes stories related to Vladimir's true feelings for Sofie, on pages 72 through 73 of *A Convergence of Lives.*

42 Women could obtain permission for travel from a father or husband or by being invited to stay with a family member or close friend. Once Kovalevskaya established a residence in Heidelberg, several other women joined her household.

43 Kochina, *Love and Mathematics*, p 53.

44 Sofia Kovalevskaya's complexion was not as fair as her sister's. She seems to be referring to her dusky complexion again in her preface to the two plays.

45 Aleksandr Nikolaevich Korkin was a Russian mathematician known for his work on partial differential equations.

https://mathshistory.st-andrews.ac.uk/Biographies/Korkin/

46 This is a reference to her sister, Anna, and her sister's husband, Victor Jaclard, who was trying to work as a French teacher. Koblitz, p 126.

47 It is not totally clear what Sofia Kovalevskaya's brother, Fedya, is doing or thinking about here, but I have entertained the thought that he is pondering the philosophy of Herbert Spencer with whom Kovalevskaya supposedly debated a woman's ability to think abstractly. (Koblitz, p 91) On reading the passages on women's rights in Spenser's *Social Statics* (1851), it seems to me that he was supportive of a woman's right to pursue and excel in any field in which she manifested ability. He makes a case for most of the restrictions placed on women having grown out of selfish desires to suppress their abilities in order for men to have mastery over them. He clearly expressed the thought that the path of arbitrary subjugation of women in society and education was inappropriate and detrimental to social progress. So, the famous debate she is supposed to have engaged in with him in London at the home of George Eliot in October of 1869, may have been more of a lively discussion than a debate between persons holding opposite views.

https://oll.libertyfund.org/title/spencer-social-statics-1851#lf0331_head_030

48 Vladimir Kovalevsky committed suicide in April of 1883, just before he would have had to face the consequences of having engaged in some troubling business transactions. He and Sofia had begun living separately and Sofia was distressed by the thought that he might not have died had she been there to prevent his action. On receiving the news of his death, she locked herself in her room and would not eat for many days. Koblitz, *A Convergence of Lives*, p 171.

49 Koblitz, *A Convergence of Lives*, p 233.

50 Koblitz, *A Convergence of Lives*, pp 235 – 236.

51 *Nihilist Girl*, by Sofya Kovalevskaya, translated by Natasha Kolchevska with Mary Zirin. p 38.

52 Koblitz, *A Convergence of Lives*, p 235 n.

53 An English translation of a letter from Weierstrass to Kovalevskaya is in Roger Cooke's *The Mathematics of Sonya Kovalevskaya* pp 98 – 99.

54 Kennedy, *Little Sparrow*, p 206.

55 Anna C. Leffler, *Sonya Kovalevskaya, a Biography*, pp 25 – 39.

56 The sky was so cloudy,

The night was so heavy,

Like our love, exactly,

Made only for tears.

57 Anna C. Leffler, *Sonya Kovalevskaya, a Biography*, p 84.

58 In *A Convergence of Lives*, Koblitz sites a letter written by Kovalevskaya in September of 1890 in which she speaks of intentions to extend the story of her childhood at least through the Heidelberg days, which is where this poem's narrative ends. p 266.

59 The apartment was located on Askanischer Platz near the *Anhalter Bahnhof*.

60 This link gives the history of the famous train station and plans for a museum to be built at the site.

https://www.berlinexperiences.com/featured-berlin-experiences/walk-through-the-ruins-of-anhalter-bahnhof/

61 Reinhard Bölling, Briefwechsel zwischen Karl Weierstraß und Sofija Kowalevskaja. pp 86 – 88.

62 This first stanza is a quotation from the end of Act One in the play *Die Verhängennisvolle Gabel* by A. von Platen.

63 Bölling, *Briefwechsel zwischen Karl Weierstraß und Sofija Kowalevskaja*, pp 352 – 359.

64 Bölling, *Briefwechsel zwischen Karl Weierstraß und Sofija Kowalevskaja*, p 356.

65 Kennedy, *Little Sparrow*, pp 68 – 70.

66 Kennedy, *Little Sparrow*, pp 263 – 278.

67 Kochina, *Love and Mathematics*, p 237.

68 Toynbee Hall, the first settlement house, was established in 1884 in London's East End by Henrietta and Samuel Barnett, so that students from Oxford and Cambridge could work as volunteers in social reform, experiencing first-hand the effects of poverty. This launched an international movement which soon spread across Europe and to the United States.

69 In Thomas Moore's poem *Paradise and the Peri*, a peri gains entrance to heaven after three attempts at finding the gift most dear to God. The first attempt is a heart that bleeds and breaks for a cause. Next is a precious sigh of pure, self-sacrificing love. The third gift, the one that gets the peri into heaven, is a tear from a repentant sinner's cheek.

70 Anna C. Leffler, *Sonya Kovalevskaya, a Biography,* pp 94 – 95.

71 Sofya Kovalevskaya, *A Russian Childhood,* p 133.

72 Sofya Kovalevskaya, *A Russian Childhood,* pp 163 – 164.

73 Sofya Kovalevskaya, *A Russian Childhood,* p 165.

74 Kennedy, *Little Sparrow,* p 259.

75 Kennedy, *Little Sparrow,* p 259.

76 Kochina, *Love and Mathematics,* p 35.

77 Sofya Kovalevskaya, *A Russian Childhood,* p 152.

78 Kennedy, *Little Sparrow,* p 261.

79 G. J. Tee, 28 June, 1976, p 124.

http://www.thebookshelf.auckland.ac.nz/docs/Maths/PDF/mathschron005-012.pdf

80 Anna C. Leffler, *Sonya Kovalevskaya, a Biography,* pp 104 – 105.

81 Anna Charlotte claims that Sofie had proposed mottos for the two plays: First, "What shall it profit a man if he gain the whole world and lose his own soul?" and Second, "He who loses his life shall save it." Anna C. Leffler, *Sonya Kovalevskaya, a Biography,* p 99.

82 Anna C. Leffler, *Sonya Kovalevskaya, a Biography,* p 98.

83 Sofia Kovalevskaya, *How It Might Have Been,* Act Three.

84 The drawing is based on a detail from Heinrich Hofmann's *Christ and the Rich Young Ruler* painted in 1889 and later purchased by John D. Rockefeller. The painting is now at Riverside Church in New York.

www.ingramcontent.com/pod-product-compliance
Lightning Source LLC
Chambersburg PA
CBHW020442130626
46549CB00001B/266